F. K. Lehner J. L. Urai (Eds.)

Aspects of Tectonic Faulting

Springer
Berlin
Heidelberg
New York
Barcelona
Hong Kong
London
Milan
Paris
Singapore
Tokyo

F. K. Lehner J. L. Urai (Eds.)

ASPECTS OF TECTONIC FAULTING

In Honour of Georg Mandl

With 88 Figures, 6 in color

Springer

Florian K. Lehner
Institute for Geology and Paleontology
University of Salzburg
Hellbrunner Strasse 34
5020 Salzburg, Austria

Janos L. Urai
Geologie – Endogene Dynamik
RWTH Aachen
Lochnerstrasse 4–20
D-52056 Aachen, Germany

Text Editing and Layout:

Wouter van der Zee
Geologie – Endogene Dynamik
RWTH Aachen
Lochnerstrasse 4–20
D-52056 Aachen, Germany

ISBN 3-540-65708-8 Springer-Verlag Berlin Heidelberg New York

Cataloging-in-Publication data applied for
Die Deutsche Bibliothek - CIP-Einheitsaufnahme
Aspects of tectonic faulting: in honour of Georg Mandl/F.K. Lehner; J.L. Urai (ed.). – Berlin; Heidelberg; New York; Barcelona; Hong Kong; London; Milan; Paris; Singapore; Tokyo: Springer, 2000
ISBN 3-540-65708-8

Cover design: design & production GmbH, Heidelberg, Germany
Cover photo: Segmented normal fault in a sand-clay layered sequence near Miri, Sarawak Malaysia. (Photo by Wouter van der Zee).
Typesetting: Camera ready by W. van der Zee

Printed on acid-free paper SPIN 10552253 32/3133as-5 4 3 2 1 0

Editors' Preface

This Volume brings together twelve contributions to a symposium held in honour of GEORG MANDL at the University of Graz, Austria on December 1–2, 1995, in the year of his 70th anniversary. It is a tribute to a formidable scientist colleague and friend and a gift of gratitude to an inspiring leader and great instigator. A man, who began as a theoretical physicist, made fundamental contributions to the theory of transport processes in porous media and the mechanics of granular materials, but in his forties turned to structural geology and the mechanics of tectonic faulting – a subject that has since remained at the center of his interests and the understanding of which was substantially advanced by Georg Mandl's work. In addressing different aspects of tectonic faulting, mostly if not entirely from a theoretician's or modeler's point of view, the contributions to this Volume reveal some of the astonishing richness of the subject, the corresponding diversity in approaches and also challenges that lie ahead. They aptly evoke the broad scientific culture brought by Georg Mandl to the study of his favourite subject, a culture he had acquired in the course of a career in a nowadays rare environment of industrial research and which interested readers will find sketched in the Biographical Note included in this Volume. As such, as well as in their own right, the papers contributed to this Festschrift should be of interest to a wider community of Earth scientists.

As Editors, we wish to thank all contributors for their participation in this project, with special appreciation of patience displayed in coping with certain 'rate-controlling steps'.

Our thanks go to Professor Eckart Wallbrecher and Professor Gunter Riedmüller and to their staff for providing financial support and for preparing and hosting the Symposium. Wouter van der Zee has produced a camera-ready copy of the book and for this we owe him a special word of gratitude. Finally, we gratefully acknowledge the generous support given to this project by the management of the Shell International Exploration & Production at Rijswijk, The Netherlands.

F. K. LEHNER J. L. URAI

photo: Bruno van Moerkerken

GEORG MANDL

Biographical Note

GEORG MANDL was born on the 27th of July 1925 in Feldkirch in Vorarlberg, Austria, as the first of the three sons of Dr. Georg and Franziska Mandl. His father was a lawyer, who came from a Viennese family and moved to Feldkirch, the home of Georg's mother. Georg and his brothers enjoyed a comfortable and protected childhood in the private quarters of a hotel inherited and managed by his mother.

Georg appears to have had an early interest in biology, but, in his own words: "no interest at all in geography and maps", a preference, it seems, for exploring the inner workings of things and little interest in mere "topographic facts". The intellectually formative years of upper high school at the 400 years old Gymnasium at Feldkirch were enriched by his close friendship with Werner Greub, the future renowned algebraist. In the exchanges with his friend, Georg appears to have recognized and defined his preference for *natural philosophy*. While his friend was to enrich its language – mathematics – he would use the same language to formulate and discover hitherto unknown relations among nature's objects. "Coming to grips with things" became Georg's challenge, beginning with the catching of bats in Feldkirch's old bell tower in a manner that is said to have been admired by his mathematical friend.

Georg's childhood and youth coincided with the dramatic second half of the short-lived first Austrian Republic and the second World war. Having earned the Matura at the natural sciences branch of the Feldkirch Gymnasium in 1943, he was drafted in 1944 to serve in a radio communication unit of the Wehrmacht. At a critical moment, he fell ill with Hepatitis and on being temporarily discharged from service, he saluted his captain with a rather daring, Austrian-style "Habe die Ehre!" (I have the honour) that left little doubt as to where he stood.

Georg entered the University of Innsbruck to study theoretical physics in the Spring of 1946, but after a semester he moved to Vienna, where H. Thirring, W. Glaser and J. Radon were his teachers in theoretical physics and mathematics. He completed his studies in 1951 and wrote his doctoral dissertation on "The Foundation of Geometrical Optics on Maxwell's Field Theory". From 1953 to 1954 Georg worked for Siemens & Halske as a project engineer for control systems, while looking for a position elsewhere in research. For a young Austrian physicist in those days this also meant to consider going abroad. The opportunity to do so arose for Georg and Berta Unterberger in 1954 on the very day of their wedding, when, on a recommendation by Professor van der Waerden of ETH Zürich, Georg was offered a position as a research physicist in the Koninklijke Shell Laboratorium Amsterdam. Georg and Berta moved to Amsterdam, for Georg the start of a long and fruitful career as a scientist with Shell Research and for the young couple the beginning of a cosmopolitan life. Holland and the open and informal way of its friendly people appealed to them. They soon founded a circle of friends for a lifetime and acquired the impeccable Dutch that Georg loves to embellish with colourful examples of its rich treasure of proverbs and sayings. He was to be outdone in this only by his daughter Barbara, who was born and raised in Holland and who shares many of her father's tastes and interests, later becoming a theoretical physicist herself.

The Koninklijke/Shell Laboratorium Amsterdam, also known as KSLA, and its mathematics and theoretical physics department, headed by Professor Oosterhof at the time of Georg Mandl's entry, went through a golden age in the fifties and early sixties. It owed its outstanding reputation to a tradition in which staff members and managers alike were accomplished scientists with close links to academia, indeed often holding professorships in nearby universities. An important area of research was the theory of transport processes in porous materials, on which Georg set out to work. Apart from his theoretical colleagues, among whom were H. Beckers, H.C. Brinckman, J.H. Kruizinga, and H.J. Merck, Georg also encountered an exceptionally strong group involved in the study of hydrocarbon recovery processes, including D.N. Dietz, L.J. Klinkenberg, P. van Meurs, C. van der Poel and others, whose work had an enormous impact on the field and advanced the practice of petroleum reservoir engineering decisively. As one of Georg's early contributions to this area, one finds – rather fittingly – a derivation of Darcy's Law from the Stokes Equations, unfortunately published only in abstract form. The actual report involved the use of the so-called 'averaging theorem' and emphasized the fact that in the course of such a 'derivation' of Darcy's Law one had to invoke certain statistical properties of a porous medium, the precise geometrical nature of which was typically left unspecified, except for assuming it to be of such a kind as to justify the neglect of certain correlation terms in the averaged, macroscopic equations of motion. In a major publication from that period, coauthored by G. Mandl and J.H. Kruizinga, the theory of immiscible two-phase flow through a porous medium was developed in a very complete manner, employing the method of characteristics, and providing experimental evidence of the predicted spontaneous appearance of saturation jumps.

In 1960 Georg and many of his colleagues moved from Amsterdam to Rijswijk near The Hague into the newly founded Koninklijke/Shell Exploratie en Produktie Laboratorium (KSEPL), which was to focus on R & D related to Shell's upstream activities. At Rijswijk, the group working on oil recovery processes under the leadership of J. van Heiningen embarked on a very comprehensive and ambitious programme of research. Scaled physical analogue experiments on model oil fields were the principal tool for studying oil recovery processes in these days. Unlike their modern digital counterparts, these analogue computers provided an excellent opportunity for young investigators to acquire physical insight and build intuition. Steam injection was considered an effective way to accelerate and enhance recovery from heavy-oil reservoirs; the 'steam drive' process was soon to be tested in the Dutch Schoonebeek field and 'steam soak' stimulation was applied on a large scale in Venezuela. These engineering developments prompted Georg to continue his collaboration with

Jan Kruizinga, now on the theory of heat and mass transport in steam-drive processes. Their work provided an elegant quantitative description of a steam zone advancing through a porous formation.

In 1963 Georg moved to the Shell Bellaire Research Laboratories in Houston, Texas, as an exchange scientist under Shell's internal 'sabbatical leave' program, which was developed to promote the communication between US and European research staff. The two years at Houston turned out to be tremendously stimulating for Georg and had a decisive influence on his career. Among other things in his luggage, he carried his work on steam drive processes and set out to improve on that theory, testing it against an impressive set of experiments that were performed by his new colleague Charley Volek. Their joint paper, published in the SPE Journal, became a milestone in the field of heat and mass transfer processes in porous media.

While still at Rijswijk, Georg had already turned to problems of coupled flow and deformation in porous geological materials. In a careful micromechanical analysis, building upon earlier work of Terzaghi and Heinrich & Desoyer and published 1964, he clarified the significance of different macroscopic stress concepts and the role – if any – of buoyancy effects in porous media. The Houston interlude could not have come at a better moment for Georg. While enjoying the academic priviledges of a member of the theoretical department headed by A. Ginzburg, he found himself next-door to John Handin's outstanding experimental rock mechanics group and, even more important, he also found an interest in quantitative models of geological structures among structural geologists, nurtured by their former director M. King Hubbert. A true pioneer of this new trend at the Bellair lab was Helmer Odé, author of a number of highly influencial papers on the theory of faulting and on folding of stratified viscoelastic layers, and collaborator on a theory of gravity instabilities in two-layer salt/overburden systems with Maurice Biot who at that time worked as a consultant for Shell. Odé had documented the application of various branches of solid mechanics to problems in structural geology in a number of reports that were written in a scholarly style. To a mature theoretician, the clear message to be read from these reports was however that comparatively little had been accomplished so far and that much more remained to be done in this field.

Helmer Odé's message appears to have been received by Georg with great enthusiasm. Field trips to the Rocky Mountains were arranged to "discover the boundary conditions". What he saw, was simply awsome to a theoretical physicists mind. Mere facts and "topography" on one hand and key observations, revealing the truth of or refuting some conjectured mechanism on the other: where was the dividing line? Obviously, field trips had to be taken only now and then, to avoid confusion ...

Upon returning to The Netherlands and KSEPL, Georg was asked by its director J. van Heiningen to set up a research section within the Geology Department, named *Geomechanics – Structural Geology*. The work of this new group was to advance the quantitative, mechanical understanding of geological processes and structures relevant to the exploration and production of hydrocarbons. This project was to become Georg Mandl's unique contribution to a now vanished industrial research culture, promoted and supported by a generation of farsighted research directors for almost twenty years until 1984, when Georg retired from his duties as section head in order to concentrate on the writing of his "Mechanics of Tectonic Faulting". This book can be read as an account of the major accomplishments of Georg's group over the years. It also puts on record a level of competence and understanding that was attained by its members and was communicated to structural geologists throughout Shell by means of a continuing education system.

The title of Georg's book reflects the early choice of a main theme for his group. In the beginning, questions concerning the theoretical foundations of a theory of frictional plasticity stood in the foreground, especially questions concerning the kinematics of frictional plastic flow which bears directly on the geometry of faults, when these are seen as velocity discontinuities. These problems were addressed in a characteristically thorough theoretical study, coauthored by R. Fernández-Luque, who had become interested in the micromechanics of granular materials. Also in this first period belongs the now classical experimental investigation, with L.N.J. de Jong and A. Maltha, into the mechanics and structural evolution of shear zones in granular materials. For the first time, a number of characteristics of tectonic shear zones were observed under controlled conditions in a large ring-shear apparatus. Focussing on the role of faults in hydrocarbon migration and trapping, the evolution of fault zone hydraulic properties was investigated experimentally in the same apparatus, while simultaneously undertaken field studies produced valuable insights into the origin of clay smears in synsedimentary faults. The frequent field trips to fault exposures that were undertaken by Georg's group in those days were usually prepared for by R. Spijer, senior and for a while only geologist in the group, and its "scientific censor". Bob Spijer's sharp and witty contributions to the ongoing process of conjecture and refutation, which tended to emphasize the second, was the kind of "needling" that Georg must have found stimulating, since he always enjoyed it and continues to do so. Bob would return from a "vacation in the Haselgebirge", to pull Georg's leg, and Georg was delighted at being taught stratigraphy in this unforgettable manner.

Next to fundamental studies directed towards a better understanding of frictional-plastic material response and related experimental and field work, the numerical modeling of tectonic faulting became a major theme of the group. Classical slip-line theory for Coulomb-plastic materials was applied in a series of papers on the theory of listric growth faults in overpressured sediments with W. Crans. Work by J.N. Thomas, J.V. Walters and H.W.M. Witlox on improved elasto-plastic finite-element schemes resulted in some of the first numerical studies of tectonic faulting as a shear localization phenomenon. Efforts in this direction by members of Georg's section provided a realistic assessment of the capabilities and potential of finite element modeling in this problem area. They also prompted a resurrection of sandbox experiments as physical analogue models of different styles of tectonic faulting. While one had to be mindful of the fact that not all criteria of similitude of model and prototype were satisfied in sandbox experiments, the close match of complex natural fault systems and experimentally obtained shearband patterns could nevertheless be explained in terms of classical slip-line theory, thus providing compelling evidence of the applicability of this type of theory to problems of tectonic faulting.

In hindsight it seems clear that the perfection of sandbox experiments in Georg's group may have produced the largest 'return on investment', for it remains a fact that thirty years of effort by the numerical modeling community have not yet produced in a satisfactory manner any of the more complex fault patterns seen in plane deformation experiments – not to speak of problems in three dimensions. Shear bands in sand were first visualized with the aid of X-ray tomography by A. Maltha. Many of the improvements in experimental technique that were introduced by W. Horsfield and other members of Georg's team subsequently found their way to a growing academic community of sandbox modelers. Richard Harkness, the accomplished soil mechanicist, also played an important role in this; he had come from Southampton University and became a close associate of Georg Mandl during the seventies and early eighties.

Questions of fault geometry, of the continuity of fault surfaces in three dimensions, and of the accumulation of tectonic strain on sets of parallel faults were addressed by Georg and some of his coworkers in the eighties in a series of papers. These demonstrate a new level of understanding that had been achieved by a consistent application of theoretical and experimental tools to the problem of faulting, seen as a velocity discontinuity in plastic deformation. The limits of this approach to tectonic faulting have been clearly spelled out by Georg himself. Ductile shear zones, which come into play when major faults transsect the brittle crust, clearly lie beyond its scope. In the light of this restriction to the brittle crust, the possible role of fracture mechanics in a theory of incipient tectonic faulting has been the subject of an ongoing debate between Georg and many of his associates and colleagues in academia. Foremost among these were Professor Neville Price and his group at Imperial College. On numerous superb field trips, Neville introduced the physicist-geologist Georg to fracture and joint systems, sometimes offering interpretations in fracture mechanics terms that were almost as enticing as the bottle of Bordeaux that would often bring their diverging views into harmony by the end of the day. Be it outcrop or lecture room - Neville Price and John Cosgrove contributed substantially over the years to the courses organized by Georg's section for Shell's structural geologists. However, an entirely satisfactory distinction between joints and faults on which everyone could agree still appears to be lacking. Is it not time to open a new bottle?

Georg Mandl's *Geomechanics – Structural Geology* group was held together and propelled by his inspired leadership. The idiosyncrasies of its members could be truly demanding at times, but they added strength and profile to an exceptionally motivated team and allowed lasting personal friendships to develop in an atmosphere of excitement and discovery. No one could certify this better than Berta, who hosted countless unforgettable evenings at Muurbloemweg 92, Den Haag, for Georg's team and for scientist colleagues from all over the world.

Although this chapter in Georg's career was to end in 1984, his flair for bringing people together on a venture of discovery and his apparently boundless energy soon enabled him to found a new circle of enthusiasts. In 1984 he began teaching a course in Geomechanics at the Institute of Engineering Geology and Applied Mineralogy of the Technical University of Graz, Austria, as well as at a number of Dutch and German universities. More properly, perhaps, the title of his course could have been "tectonomechanics", a term that had been used by Leopold Müller, the famous rock mechanicist and friend and supporter of Georg Mandl. It aptly characterizes the main thrust of Georg's book, and he had already used it in his Gauss Medal Lecture for Leopold Müller, entitled *Tectonomechanics – Cinderella of Geology?* After his retirement from KSEPL in 1986, and having returned to Austria, Georg developed closer ties to the Geology Department at Graz University and soon succeeded initiating the Graz *Tectonomechanics Colloquium*, a continuing series of semiannual two-day meetings organized jointly by Professor E. Wallbrecher and Professor G. Riedmüller, heads of the Institute for Geology and Paleantology of the University of Graz and of the Institute of Engineering Geology and Applied Mineralogy at the TU Graz, respectively. These very informal meetings have brought together, for more than ten years now, an international group of enthusiasts of Georg's Cinderella, for the presentation and discussion of new developments in tectonomechanics and geomechanics. Georg himself continues to be a regular contributor to these meetings, animator of lively exchanges, and inspiring young and old by his enthusiastic response to new ideas. It is at such occasions, that his ability to seize on the positive and constructive part of an idea, argument, or observation, shines as one of his most precious qualities. A wonderful demonstration of this gift was his delighted reaction to one particular review of his book, in which the reviewer had praise for Georg Mandl the geologist, but had questioned his competence when it came to physics.

In 1990, Georg Mandl received his *venia docendi* upon having submitted his inaugural dissertation, entitled "Mechanik tektonischer Störungen" to the Technical University of Graz and, in 1995, he received an Honorary Professorship for Geomechanics by the TU Graz, where he continues to lecture. Those who have followed his lectures must at times have marvelled at his ability to reduce the complex phenomenology of tectonic faults to but a few lines and points in a Mohr stress plane. Out of this came the suggestion to write another book, that would condense a wealth of insights and demonstrations into "something simple and concise". Georg has taken up this suggestion and, in the meantime has completed his second book. The friends and colleagues of Georg Mandl look forward to it as they do to his indispensable presence at many more Tectonomechanics Colloquia.

FLORIAN LEHNER

List of Publications by Georg Mandl

Journal Articles

Zur Begründung der Strahlenoptik aus der Maxwell'schen Feldtheorie. *Acta Physica Austriaca*, Bd. VII, Heft 4, 365–389, 1953.

(With J.H. Kruizinga) Über das spontane Auftreten von Unstetigkeiten bei der schleichenden Strömung zweier Flüssigkeiten durch ein poröses Medium. *Z.f. angew. Physik*, **13**/2, 81–90, 1961.

Zur statistischen Begründung der Mechanik flüssigkeitsgefüllter Medien. *Progr. wissensch. Jahrestagung Ges. angew. Math. & Mech.* (GAMM), Technische Hochschule Karlsruhe, 1963.

Change in skeletal volume of a fluid-filled porous body under stress. *J. Mech. Phys. Solids*, **12**, 299–307, 1964.

(With C.W. Volek) Heat and mass transport in steam-drive processes. *Trans. AIME*, **246**, 59–79, 1969.

(With R. Fernández Luque) Fully developed plastic shear flow of granular materials. *Géotechnique*, **20**, 277–307, 1970.

(With C. Crans and G. Shippam) *Geomechanische Modelle tektonischer Strukturen.* SFB-77 Jahresbericht 1973, University of Karlsruhe, Germany, 1973.

(With C. Kruit) Sedimentary squeeze structures indicative of paleoslope. *Coloquio de Estratigr. Paleogeogr. del Cretacio de Españia, Trab. Congr. Reun.*, **7**(1), 93–101, 1975.

(With L.N.J. de Jong and A. Maltha) Shear zones in granular materials: An experimental study of their structure and genesis. *Rock Mechanics*, **9**, 95–144, 1977.

(With K.J. Weber, F.K. Lehner, W.F. Pilaar and R.G. Precious) The role of faults in hydrocarbon migration and trapping in Nigerian growth fault structures. *Proc. 10th Annual Offshore Technology Conference*, Houston, Texas, Vol. 4, 2643–2653, 1978.

(With W. Crans and J. Haremboure) On the theory of growth faulting: A geomechanical delta model based on gravity sliding. *J. Petrol. Geol.*, **2**(3), 265–307, 1980.

(With W. Crans) On the theory of growth faulting, Part IIa. *J. Petrol. Geol.*, **3**(2), 209–236, 1980.

(With W. Crans) On the theory of growth faulting, Part IIb. *J. Petrol. Geol.*, **3**(3), 333–355, 1981.

(With W. Crans) On the theory of growth faulting, Part IIc. *J. Petrol. Geol.*, **3**(4), 455–476, 1981.

(With W. Crans) Gravitational gliding in deltas. In: *Thrust and Nappe Tectonics*, edited by K.R. McClay and N. Price. Geol. Soc. London Spec. Publ. No. 9, pp. 41–54, 1981.

(With G. Shippam) Mechanical model of thrust sheet gliding and imbrication. In: *Thrust and Nappe Tectonics*, edited by K.R. McClay and N. Price. Geol. Soc. London Spec. Publ. No. 9, pp. 79–97, 1981.

Tektonomechanik – Stiefkind der Geologie? *Jahrb. Braunschweigische wiss. Ges.*, 27–51, 1983.

(With M.A. Naylor and C.H.K. Sijpestein) Fault geometries in basement-induced wrench faulting under different initial stress states. *J. Struct. Geol.*, **8**(7), 737–752, 1986.

Discontinuous fault zones. *J. Struct. Geol.*, **9**(1), 105–110, 1987.

Tectonic deformation by rotating parallel faults: The "bookshelf" mechanism. *Tectonophysics*, **141**, 277–316, 1987.

Hydrocarbon migration by hydraulic fracturing. In: *Deformation of Sediments and Sedimentary Rocks*, edited by M.E. Jones and R.M.F. Preston. Geol. Soc. London Spec. Publ. No. 29, pp. 39–53, 1987.

Modelling incipient tectonic faulting in the brittle crust. In: *Mechanics of Jointed and Faulted Rocks*, edited by H.-P. Rossmanith. Balkema, Rotterdam, pp. 29–40, 1990.

Paläomechanische Gebirgsparameter – die grossen Unbekannten bei der Nutzung der Erdkruste. *Berg- und Hüttenmännische Monatshefte*, **4**/91, 129–134, 1991.

Books

Mechanics of Tectonic Faulting, XIV+407 pp. Elsevier, Amsterdam, 1988; reprinted 1993.

Faulting in Brittle Rocks, X+434 pp. Springer-Verlag, Berlin Heidelberg 2000.

Table of Contents

Contributors

Dr. Vincent Auzias Laboratoire de Géophysique et Tectonique, Université de Montpellier II, Place E. Bataillon, 34095 Montpellier Cédex 5, Case 060, France.

Dr. Peter Cundall PO Box 195 (231 2nd Street), Marine on St. Croix, Minnesota 55047, USA.

Dr. Johann Genser Institute for Geology and Paleontology, University of Salzburg, Hellbrunner Strasse 34, 5020 Salzburg, Austria.

Maike Knoop Shell International Exploration and Production, Volmerlaan 8, 2288-GD Rijswijk, The Netherlands.

Dr. Walter Kurz Institute for Geology and Paleontology, University of Salzburg, Hellbrunner Strasse 34, 5020 Salzburg, Austria.

Dr. Florian K. Lehner Institute for Geology and Paleontology, University of Salzburg, Hellbrunner Strasse 34, 5020 Salzburg, Austria.

Dr. Wolfgang Lenhardt ZAMG-Central Institute for Meteorology and Geodynamics, Hohe Warte 38, 1191 Vienna, Austria.

Dr. Yves Leroy Laboratoire de Mécanique des Solides, École Polytechnique, URA CNRS 317, 91128 Palaiseau Cedex, France.

Dr. Franz Nemes Institute for Geology and Paleontology, University of Salzburg, Hellbrunner Strasse 34, 5020 Salzburg, Austria.

Professor Dr. Franz Neubauer Institute for Geology and Paleontology, University of Salzburg, Hellbrunner Strasse 34, 5020 Salzburg, Austria.

Professor Dr. Horst Neugebauer Institute for Geodynamics, University of Bonn, Nussallee 8, 53115 Bonn Germany.

Dr. Dick Nieuwland Dept. of Earth Sciences, Vrije Universiteit, De Boelelaan 1081, 1085 Amsterdam, The Netherlands.

Mag. Eike Paul Institute for Geology and Paleontology, University of Salzburg, Hellbrunner Strasse 34, 5020 Salzburg, Austria.

Professor Dr. Jean-Pierre Petit Laboratoire de Géophysique et Tectonique, Université de Montpellier II, Place E. Bataillon, 34095 Montpellier Cédex 5, Case 060, France.

Professor Dr. David D. Pollard Department of Geological and Environmental Sciences, Stanford University, Stanford, CA 94305-2115, USA.

Dr. Keith Rawnsley Elf-Aquitaine Production, Pau, France.

Dr. Thierry Rives Elf-Aquitaine Production, Pau, France.

Univ. Dozent Dr. Hans-Peter Rossmanith Institute for Mechanics, Technical University of Vienna, Wiedner Hauptstrasse 8–10/325, 1040 Vienna, Austria.

Dr. William Sassi Institute Français du Pétrole, 1&4, av. de Bois-Préau BP 311, 92506 Rueil Malmaison Cedex, France.

Professor Dr. Nicolas Triantafyllidis Department of Aerospace Engineering, The University of Michigan Ann Arbor, Ann Arbor, Michigan 48109-2140, USA.

Dr. Koji Uenishi Institute for Mechanics, Technical University of Vienna, Wiedner Hauptstrasse 8–10/325, 1040 Vienna, Austria.

Dr. Wolfgang Unzog Institute for Geology and Paleontology, University of Graz, Heinrichstrasse 26, 8010 Graz, Austria.

Professor Dr. Janos Urai Institut für Geologie, RWTH Aachen, Lochnerstrasse 4–20, Aachen, Germany.

Professor Dr. Eckart Wallbrecher Institute for Geology and Paleontology, University of Graz, Heinrichstrasse 26, 8010 Graz, Austria.

Dr. Xianda Wang Institute for Geology and Paleontology, University of Salzburg, Hellbrunner Strasse 34, 5020 Salzburg, Austria.

Dr. Emanuel Willemse Shell International Exploration and Production, Volmerlaan 8, 2288-GD Rijswijk, The Netherlands.

Mag. Ernst Willingshofer Institute for Geology and Paleontology, University of Graz, Heinrichstrasse 26, 8010 Graz, Austria.

Numerical experiments on rough joints in shear using a bonded particle model

Peter A. Cundall

Itasca Consulting Group, Inc. Minneapolis, Minnesota, USA.

1 Introduction and background

Rock joints and rough faults exhibit two prominent characteristics. First, the relation between peak shear strength and normal stress is nonlinear; second, the peak dilatation angle is strongly dependent on normal stress. The accepted explanation for the dependence on normal stress relies on the influence of surface roughness. At low normal stress, the two sides of the joint "ride up" over each other with minimal damage to asperities, giving rise to a high dilatation angle and a high apparent-friction angle. At high normal stress, some asperities fail because their strengths are comparable to local stresses developed near contact points. The apparent friction and dilatation angles are therefore lower, as less "riding-up" occurs.

A preliminary study is reported here in which the microstructure of rock is represented explicitly in numerical simulations as a synthetic solid consisting of discs bonded in the shear and normal directions at all contacts. Because each element of the microstructure can fail, the effects of different normal stresses are accommodated automatically. A somewhat related study was performed by Mora & Place (1994), using circular particles with elastic-brittle bonds that are active in the normal direction of contacts only. However, the objective was to investigate slip-stick instabilities rather than the quasi-static dependence of strength on normal stress. There have also been analytical formulations for deformation, based on the Hertz theory for contact between opposing spherical asperities (e.g., Swan & Zongqi, 1985). Idealized failure modes were assumed by Liu (1996) to propose a semi-analytical model of shear failure of rough joints. In contrast, the work described here is an attempt to allow the failure processes adjacent to the joint surfaces to develop automatically, without prescribing any potential locations of cracks. We make no attempt to reproduce the elastic properties of rock joints. The simulations are performed in two dimensions, although the extension to three dimensions is trivial (but time-consuming).

2 Simulation of rock with bonded particles

Rock is a heterogeneous material that can develop numerous microcracks during the loading process. By modeling a rock sample as a collection of separate particles bonded together at their contact points, the simulated material can develop "microcracks" as individual bonds break when the sample is loaded.

Larger scale cracks develop naturally as microcracks coalesce into bands or link up into contiguous strings. The bonded-particle approach is described in some detail by Potyondy *et al.* (1996) and compared to similar methods that idealize solid material as lattices of structural units, such as springs, fuses or particles. In the simulations described here, a dense packing of circular particles is created and modeled with the distinct element method (Cundall & Strack, 1979), which is an explicit, time-marching solution to the equations of motion. Although particles are assumed to be rigid, deformability of the assembly derives from normal and shear springs that are used to represent compliant contacts between particles. Each spring also has a strength that represents intact bonding: when either a tensile normal- or shear-force limit is reached, the bond breaks and carries no tension thereafter. However the contact can still support a compressive normal force and a shear force of magnitude less than μF_n, where μ is the friction coefficient and F_n is the normal force (positive for compression). Damping is incorporated in the equations of motion by applying a damping force, F_d, to each degree-of-freedom, given by

$$F_d = -\alpha |F_u| \mathrm{sgn}(\dot{u}) \tag{1}$$

where α is a non-dimensional damping factor (set to 0.7), F_u is the algebraic sum of forces acting on the degree-of-freedom and $\dot{u}$ is its velocity. It can be shown (Itasca, 1995) that Equation (1) corresponds to hysteretic damping of ratio D to critical damping, where $\pi D = \alpha$.

The formulation summarized above is embodied in a computer program called PFC^{2D} (Particle Flow Code, Itasca, 1995). The micro-properties used in PFC^{2D} may be chosen through a calibration procedure in which the behavior of a numerical sample is matched to that of a physical sample under laboratory conditions.

3 Creation of a shear box containing a rough joint

Particles are packed at random into a box of dimension 20 units by 2 units, creating a sample consisting of 10,000 disks. In order to avoid crystalline packings, particle radii are chosen at random within a specified size distribution and are uniformly distributed between 0.025 to 0.04 size units. A packing algorithm is used that simultaneously ensures low initial contact forces (in relation to forces induced by episodes of loading) and no disconnected particles.

After packing is complete and the assembly is in equilibrium, the confining boundaries are removed and the set of boundary particles identified. This set is divided into two groups, separated by the horizontal centerline of the assembly, to form the top and bottom containers of a simple shear-box. Before performing the shear test, all contacts are bonded, except those corresponding to the "joint" to be tested. In the tests reported here, the joint is given a roughness profile consisting of the sum of two periodic waves, as follows:

$$y = 0.75\{0.15 \sin(20\pi x/w) + 0.1 \sin(35\pi(x + 1.1)/w)\} \tag{2}$$

where y is the deviation from the centerline, x is the distance from the left side of the shear box, and w is the width of the shear box. The amplitude of the joint deviation decays linearly to zero within a distance of 4 units from either end of the shear box, to produce smooth entry and exit conditions. Contacts within a distance of 0.08 units on either side of the joint track, as given by Equation (2), are left unbonded; all other contacts are given equal shear and normal bond strengths of 10^6 force units. Those contacts adjacent to the top and bottom edges of the shear box are given strengths increased by a factor of ten, in order to prevent spurious failure at the boundaries. Shear and normal contact stiffnesses are linear, with spring constants equal to 0.7×10^9 units (force divided by displacement). An overall view of the assembly is shown in Figure 1, with boundary particles colored black and unbonded particles colored in red; the remaining particles are colored green. A close-up of the left-hand end of the assembly is illustrated in Figure 2, with bonds shown as black lines joining particle centers. The band of unbonded joint material is visible as the area without bonds.

4 Test procedure

If a shear test is performed on an unbonded, granular assembly consisting of circular, *frictionless* particles, a finite value for the overall friction coefficient is observed. The shear resistance derives from interlocking particles that must ride up in order to move past each other; energy "dissipation" occurs because strain energy is converted to kinetic energy when particles suddenly jump over one another. (The resulting kinetic motion is chaotic and is absorbed by damping in the numerical simulations). In order to determine the apparent "base" friction coefficient, a preliminary test is done with zero joint roughness, zero friction coefficient and infinite bond strength remote from the joint area. It is important for the band representing the joint to have a width of several particle diameters; otherwise, opposing particles on either side of the track may develop very large resisting forces (if the line joining their centers lies almost along the mean track and both particles are bonded to their remaining neighbors). When the joint has a thickness of ± 0.08 units, the base friction angle is found to be approximately 15° under the test conditions given previously. It is commonly observed that smooth rock surfaces have a friction angle of 30°. Consequently, the contact friction angle in the *PFC*^{2D} simulations is set to 15° in order to obtain, approximately, the desired friction angle for the ensemble.

A further calibration is performed. When expressing the results of shear tests on rock joints, the normal stress is usually normalized to the unconfined compressive strength of intact rock. In order to determine this strength for the numerical sample, the compacted assembly used for the shear tests is also used in a uniaxial test by dividing the sample into nine square sub-samples and applying rigid vertical motion to the top faces of all sub-samples simultaneously (see Figure 3, in which the controlled particles are shown in black). By averaging the vertical reaction forces, the peak stress (i.e., the unconfined compressive

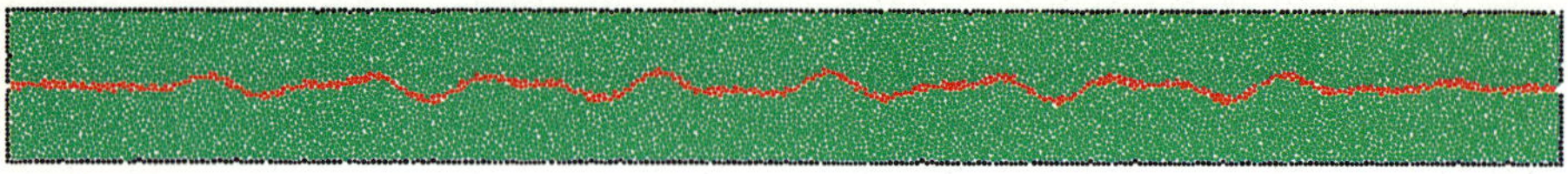

Fig. 1. Initial configuration of shear box. The upper and lower sets of controlled particles are shown in black and the set of unbonded particles (representing the joint) is shown in red.

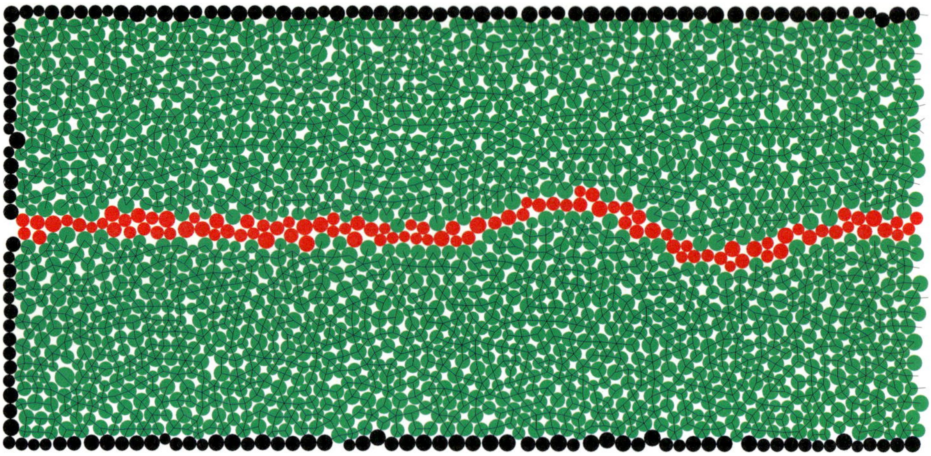

Fig. 2. Close-up of the left-hand end of the shear box. Contact bonds are shown as black lines joining particle centroids.

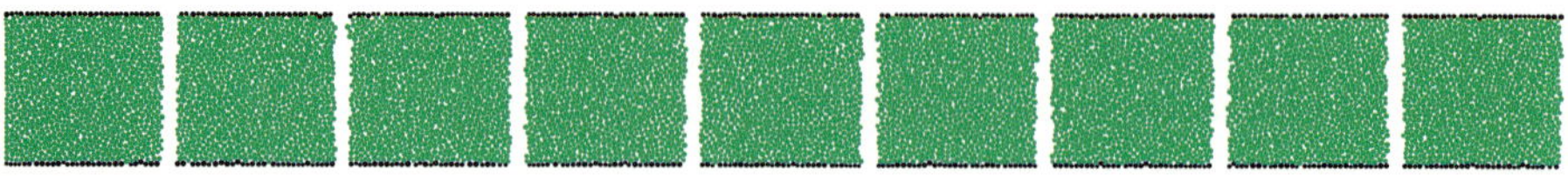

Fig. 3. Configuration for unconfined compression test: nine samples (extracted from the shear sample of Fig. 1) are tested simultaneously. Platens are indicated by black particles.

strength, σ_c) is determined to be 2×10^7 units (force divided by length, where the out-of-plane dimension is taken to be unity). For this test, all contacts are bonded with the same bond strength used for the shear-box test.

The two groups of boundary particles comprising the bottom and top of the shear box are controlled to perform the shear test. The bottom part of the box is held fixed throughout the test, and the top part is translated as a rigid unit with constant velocity in the horizontal direction. Both the vertical and horizontal reaction forces on the top part are evaluated continuously. The vertical motion is controlled by a numerical servomechanism so as to keep the vertical reaction force constant at some specified value.

5 Test results

Shear tests are performed under the conditions noted above, with various values of normal load, expressed as an average stress, σ_n, by dividing the vertical reaction force by the length of the shear box (20 units). A typical plot of shear stress (normalized to σ_c) versus shear displacement is shown in Figure 4, and a plot of normal displacement versus shear displacement is shown in Figure 5. These plots are for the case of $\sigma_n/\sigma_c = 0.65$. During the course of the test, all bond breaks are recorded. Figure 6 shows these "microcracks" for the case of

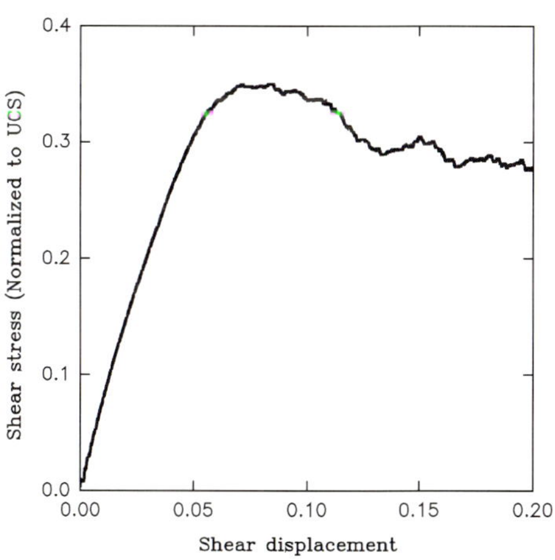

Fig. 4. Measured shear stress versus shear displacement for a normal stress of 0.65 (normalized to unconfined compressive strength).

$\sigma_n/\sigma_c = 0.25$, plotted as lines drawn perpendicular to the vectors joining the corresponding pairs of particles. The lines are of lengths equal to the distances between particle centroids. Tension cracks are shown in red and shear cracks are shown in green. The upper and lower sets of controlled particles are also illus-

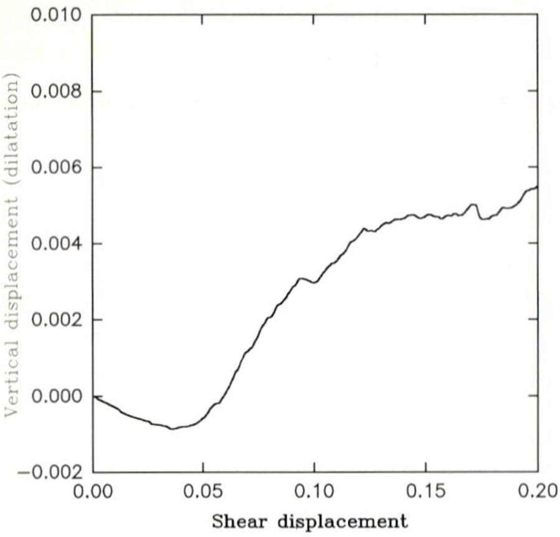

Fig. 5. Measured normal displacement versus shear displacement for a normal stress of 0.65 (normalized to unconfined compressive strength).

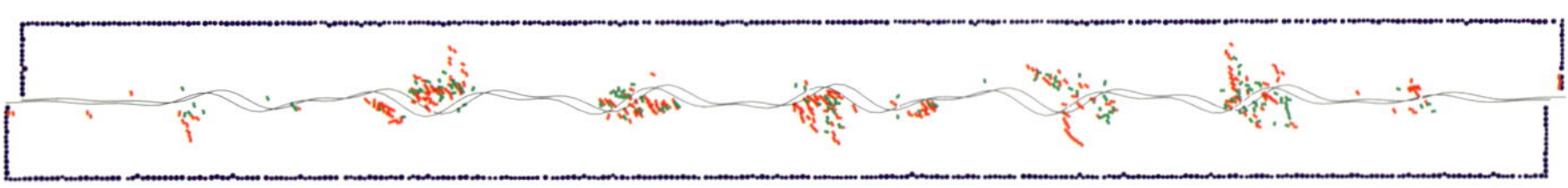

Fig. 6. Microcracks resulting from a shear displacement of 0.2 units: red denotes tension failure; green denotes shear failure. The normal stress is 0.25, normalized to UCS.

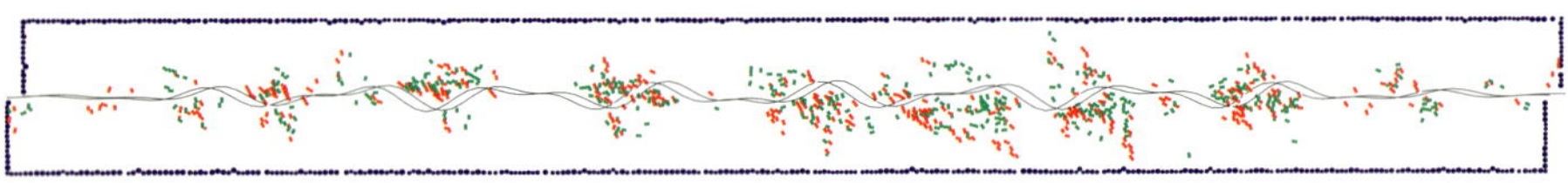

Fig. 7. Microcracks resulting from a shear displacement of 0.2 units: red denotes tension failure; green denotes shear failure. The normal stress is 0.5, normalized to UCS.

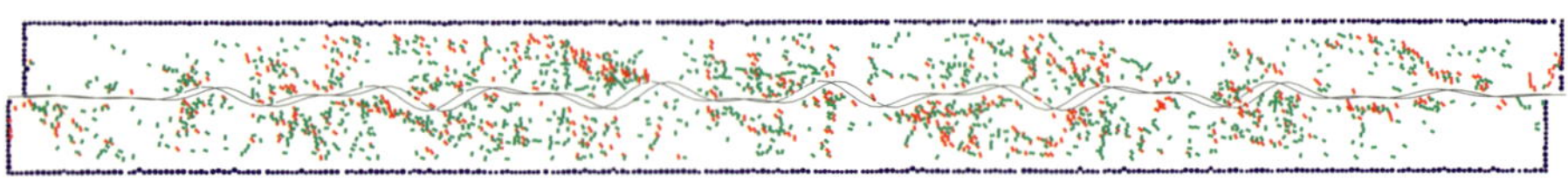

Fig. 8. Microcracks resulting from a shear displacement of 0.2 units: red denotes tension failure; green denotes shear failure. The normal stress is 0.875, normalized to UCS.

trated. Similar plots are provided in Figures 7 and 8 for the cases $\sigma_n/\sigma_c = 0.5$ and $\sigma_n/\sigma_c = 0.875$, respectively.

It is evident that damage locations are related to locations of joint asperities, and that tension cracking predominates at low normal stresses, with some tension cracks extending quite far into the intact rock. At high normal stress, the failure is much more extensive, with shear failure being dominant. There is also some suggestion that Riedel shear zones are developing.

Empirical relations proposed by Barton (e.g., see Barton & Bandis, 1990) have been in use for many years for estimating the shear strength and dilatation of rock joints. The scheme — known as the JRC/JCS model — depends on three parameters: JRC, the "joint roughness coefficient" (dimensionless), JCS, the "joint compressive strength" (in stress units) and ϕ_r, the residual friction angle in degrees. The formula for shear stress is:

$$\tau = \sigma_n \tan\{\mathrm{JRC} \log(\mathrm{JCS}/\sigma_n) + \phi_r\} \tag{3}$$

and for peak dilatation angle (in degrees) is:

$$d_n = \mathrm{JRC} \log(\mathrm{JCS}/\sigma_n) \tag{4}$$

Barton & Bandis point out that a JRC value of 20 corresponds to very rough joints and that JCS is equal to the unconfined compressive strength for unweathered joints. The shear and dilatation results from several numerical simulations are compared to predictions by the JRC/JCS model in Figures 9 and 10, respectively, using parameters $\phi_r = 24°$, JRC $= 20$ and JCS $= \sigma_c$ in Equations (3) and (4). Numerical results are denoted by crosses and the values derived from equations (3) and (4) are shown as solid lines. Shear and normal stresses are normalized by dividing by σ_c, as determined by the uniaxial compression test described previously. It appears that the numerical results conform to the general trends of the JRC/JCS model, but the computed dilatation values are 2 to 5 degrees higher than those "predicted" by the JRC/JCS model. Note that the measured dilatation values represent the maximum secant-slope of the vertical displacement versus shear displacement curve, taken over all chords equal to 15% of the maximum shear displacement.

6 Conclusions

The representation of a rough joint in rock by a micromechanical model leads to simulated behavior that is similar to that observed in real joints. In particular, the dependencies of peak shear strength and dilatation on normal stress conform reasonably well to laboratory findings and to the empirical JRC/JCS model. Further, the extent of damage corresponds with commonly-accepted notions, i.e., progressively more asperity damage occurs as the normal stress is increased. One surprising observation is that some tension cracks migrate into the solid rock at low levels of normal stress, in addition to localized damage near the joint surface. As expected, there is extensive damage at high normal stresses.

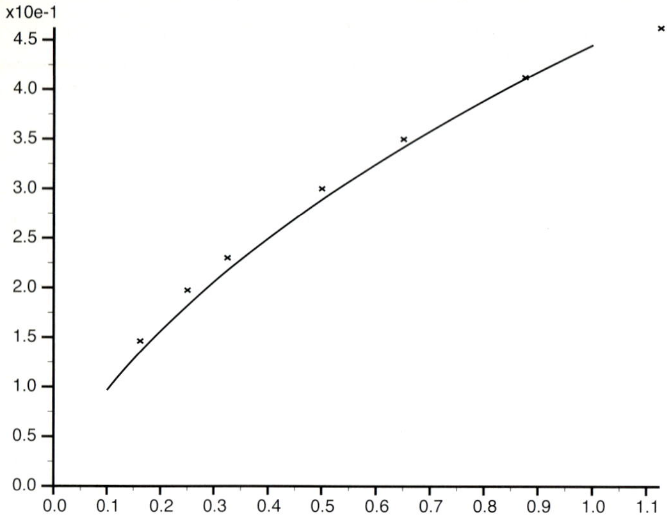

Fig. 9. Normalized shear strength versus normalized normal stress. Crosses are numerical results; solid line is from Equation (3).

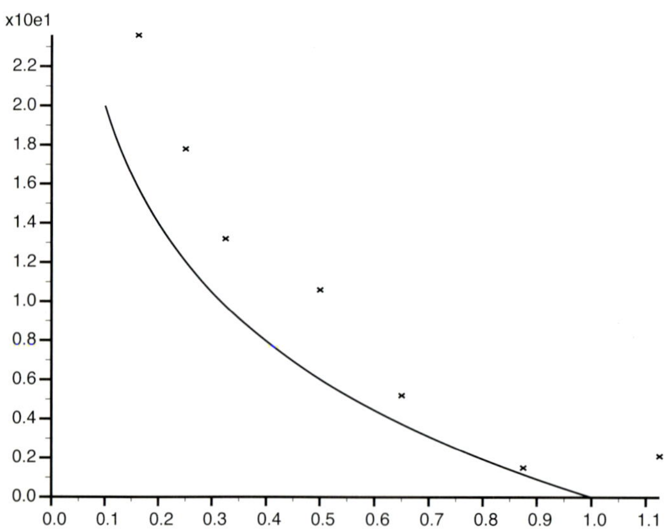

Fig. 10. Dilatation angle (in degrees) versus normalized normal stress. Crosses are numerical results; solid line is from Equation (4).

If the numerical model is accepted as valid, then it can be used to study some fundamental questions. For example, we might resolve the controversial questions of whether there is a size effect and what it depends on. The model could also be used to study the effect of various roughness *spectra* on joint behavior. There are many applications that would benefit from a better knowledge of joint mechanics. Fluid conductivity and storage is critically dependent on details of joint geometry and behavior in the normal and shear directions, with applications in high-level waste disposal and oil reservoir engineering. The effects of localized locking and rupture on earthquake source mechanisms is another field that would benefit from a better knowledge of the micro-behavior of rough faults. Finally, rock joints are a fundamental component of any engineering design of caverns, foundations and abutments in rock.

References

Barton, N. and S.Bandis (1990) Review of predictive capabilities of JRC-JCS model in engineering practice. *Rock Joints*, Barton & Stephanson (eds.), Balkema, Rotterdam.

Cundall, P.A., and O.D.L.Strack (1979) A discrete numerical model for granular assemblies. Géotechnique **29**(1):47-65.

Itasca Consulting Group, Inc. (1995) *PFC*^{2D} Manual, Version 1.1. Minneapolis, Minnesota, USA.

Liu, H. (1996) Numerical modeling of shear behavior of rock joints using microscopic structures of joint surfaces. Ph.D. thesis, Dept. of Civil Engineering, University of Minnesota.

Mora, P. and D.Place (1994) Simulation of the frictional stick-slip instability. Pure Appl. Geophys. **143**:61-87.

Potyondy, D.O., P.A.Cundall and C.A.Lee (1996) Modeling rock using bonded assemblies of circular particles. 2nd. North American Rock Mechanics Symposium, Québec, Canada.

Swan, G. and S.Zongqi (1985) Prediction of shear behavior of joints using profiles. Rock Mechanics and Rock Engineering **18**:183-212.

On the competition of interplate shear and surface erosion at convergent plate boundaries

Gerd Gottschalk and Horst J. Neugebauer

Geodynamics – Physics of the Lithosphere, University of Bonn, Germany
Tel. 0228 – 737429-30 – Fax: 0228 – 732508 – eMail: neugb@geo.uni–bonn.de

Abstract. Convergent plate boundary tectonics is characterised by a variety of deformation patterns from subduction orogenesis to continental collision zones. These tectonic features are associated with the long–term deformation of the lithospheric wedge overriding the subducting lithospheric slab. Two sets of forces appear to be predominant in the process of tectonic structure formation: the shear forces corresponding to relative plate motion along the tilted bottom of the lithospheric wedge and the changing load pattern along the free surface of the lithosphere caused by erosion and redeposition. According to the large scales involved in space and time one might argue in terms of an induced mass flow in the lithosphere driven by interplate shear, topographic erosion and isostatic readjustment.

We study the cooperation of these well established phenomena by means of a scale consistent numerical finite element approach. In order to represent the overall mechanical behaviour in various scales appropriately, we adopted an elasto–visco–plastic constitutive equation for the strain rate. It incorporates a yield criterion as well as power law creep depending on thermal energy. The long–term mass flow is traced by a regridding technique of the numerical scheme. The cooperation of subduction–driven tectonic erosion and topography–driven surface erosion leads to a rather stable structural pattern of deformation. Broad topographic uplift is reflected by a dynamically formed root–zone. This orogenic kernel is coincidently the location of exhumation of deep seated matter at the surface while in the neighbourhood on both sides fore– and back–arc or fore– and hinterland basins are formed. The hinterland basin is a result of thinning by lateral mass flow beyond the orogenic section into a hinterland plateau. Our numerical results suggest a competitive influence between both tectonic and topographic erosion. The influence of the tilt angle of the shear interface is of comparable importance for the formation of the dominant features during the long–term process. The relative strength of the lower crustal layer compared to the upper mantle structure controls the lateral distance of the induced mass flow pattern within the lithospheric wedge. Altogether, the extremely simplified scale consistent approach provides a new insight into the complex nonlinear tectonics of convergent plate boundaries.

1 Introduction

The complex physics of convergent plate boundary dynamics has been investigated so far by two major alternative concepts. In the critical wedge theory the process of tectonic accretion or erosion of a frictional Coulomb–material in response to basal shear is considered, Davis et al (1984), Dahlen (1984), Lallemand et al (1994). The approach is mainly applied to accretionary wedges of sediments at convergent plate boundaries. It was adopted for the understanding of orogenic structures and the quantification of analogue experiments by Gutscher et al (1998). The topography of the wedge is in a critical equilibrium with the loss or increase of matter at the sheared baseline. The adjustment to the equilibrium topography is achieved by the mechanical response of the Coulomb wedge. This concept favours the predominance of the rheology. Alternatively, crustal deformation in compressive orogenesis was studied with reference to visco–plastic mechanics, Beaumont et al (1994). The model concept is specialised to horizontal shear at the base of the crust in combination with a fixed stagnation point which terminates the shear and starts a downward pull. The stagnation point causes a forced squeezing of the upper layer. Thus the existence and position of the stagnation point controls the proportion between subducted and folded structures. The mechanics incorporates visco–plastic flow and faulting of the crust overlying on elastic mantle.

We propose an alternative concept which is based on the existence of large–scale interplate shear which is expressed in the seismotectonic observations, Tichelaar and Ruff (1993). Considering long–term behaviour, the shear strength of the tilted fault zone reflects the rheology of the adjacent overriding lithosphere beside the aspects of frictional faulting in the upper section, Chester (1995), Byerlee (1978). The mechanical constraints incorporate the nonelastic deformation of the mantle. Our approach is rather simple and straightforward, details are given in Gottschalk (1997). The interplate shear concept allows model experiments on orogenic tectonics through subduction. The level of specific restrictions and boundary constraints is comparatively low. According to our physical concept of the numerical approximation we raise the following questions: Is the internal mass flow induced by interplate shear during subduction sufficiently strong to play a predominant role in the development of the extended pattern of convergent plate boundary tectonics?

How far could climatic erosion be a complementary or even required outer counterpart to the shear driven deformation pattern? And finally, do consistent parameter sets exist for the large scale approach, which meet constraints coming from other independent investigations?

2 Statement of the model

In an elongated lithospheric wedge with a tilted shear boundary on the left and a horizontally fixed border on the right we let the model surface free. Mechanically a crustal and an upper mantle region is distinguished. The upper crust on the right shows an elastic backstop, Figure 1. There are three principle categories of forces acting onto this simplified structure during ongoing subduction. The depth–dependent shear along the tilted boundary causes a displacement field which deforms the crust–mantle boundary and the free surface, while isostatic readjustment acts on both interfaces. Finally the free surface undergoes a mass redistribution in response to topography–induced erosion and deposition. The induced changes in the force system drive an ongoing process of combined internal and external mass flow.

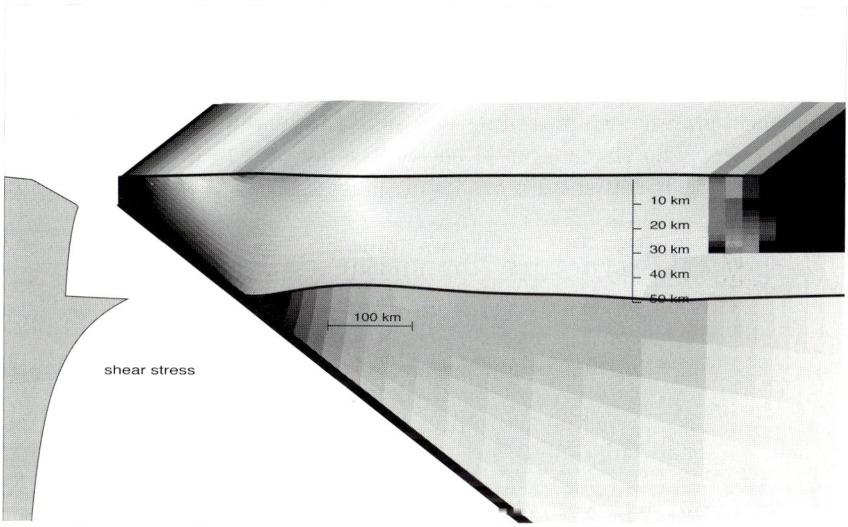

Fig. 1. Shape and dimension of the lithospheric wedge model with shear force distribution on the left. Shaded background indicates induced maximum shear stress, values are below 10^8 Pa. Beside the crust and upper mantle structure the elastic backstop is seen on the far right end.

Fault–related shear forces reflect, in a long–term average, the rock strength properties, Chester (1995). That means the local situation depends on the physical state variables, frictional properties, failure stress, fluidity, water content, elastic constants and activation energy. We assume an average frictional force which is constant in time. The force distribution varies with depth according to the prescribed rheology and is shown in Figure 1 at the left. The shear forces are calculated under the conditions:

$$\sigma_{\text{shear}} = \min\left[\sigma_{\text{plast}}(z,\dot{\varepsilon}), \sigma_{\text{frict.}}(z)\right]$$
$$\dot{\varepsilon} = \text{const} \leq 10^{-14}$$

The elasto–viscoplastic strain rate is derived from a power law creep equation combined with the von Mises yield criterion, following the above shear stress depth relation for a constant strain rate of $\dot{\varepsilon} = 10^{-14}$. The maximum shear stress which can be transfered by the shear zone are constrained by estimations following Byerlee (1978). A more detailed discussion is given in Gottschalk (1997). As a matter of consistency of model scales the erosion–deposition procedure follows a diffusion equation for the topographic height $h(x,t)$. The diffusion constant D becomes a measure for the rate of erosion in the model. The erosion rate itself is a local variable which depends on the gradient of the topography. The accumulated topographic changes are transferred to an equivalent sequential load distribution with reference to the crustal rock density. Finally the entire mass–flow within and on top of the lithospheric wedge is balanced isostatically by an Airy–type compensation. The mass is conserved in total during the deformation process. Figure 1 gives a snapshot of the calculated shear stress distribution in response to the set of body and shear forces. The basal shear along the tilted boundary is well reflected. The stress distribution reflects already the mass movements shown later on. The elastic crustal backstop is an element of low influence on the stress field, it is rather a limitation of the flow regime.

3 Large scale structure formation

On a geologically relevant time scale in the order of 10 Ma the modelled lithospheric wedge deforms into a typical structural pattern shown in Figure 2. The induced mass flow leads to a substantial thickening of the crustal layer to an orogenic swell. While tectonic erosion is feeding the root zone, ascending matter forms a topographic uplift with a window of exposed rocks from a deep origin. Further to the right an expressed thinning of the crust is seen. As Figure 2 indicates, this formation is achieved dynamically by the splitting of the mass flow from the shear boundary into an exhumation upstream and a lateral component. The latter forms a plateau–like structure of moderate thickness far right. Selected sequential steps of mass elements (black–original position, grey–deformed position) emphasize the induced movement. The modelled flow rates are in the order of mm/a. Sediments are accumulated along both sides of the topographic uplift in the so called fore– and hinterland depressions. Especially the tip zone of the wedge accommodates deposited material which becomes part of a downstream close to the shear interface. Thereafter part of the exhumation upstream turns over into the frontal downstream, forming a megafold, which is presented in detail later on. In summary the interplate shear forces along the subduction interface evolve a significant pattern of structure formation within the entire lithospheric wedge.

4 Critical control of structure formation

A quantitative comparison of model families allows the evaluation of critical parameters of the system. In the following models the tilted shear forces and the constitutive equation for the strain rate are kept constant. With this reference we discuss alterations of the topographic erosion and implications of the dip angle

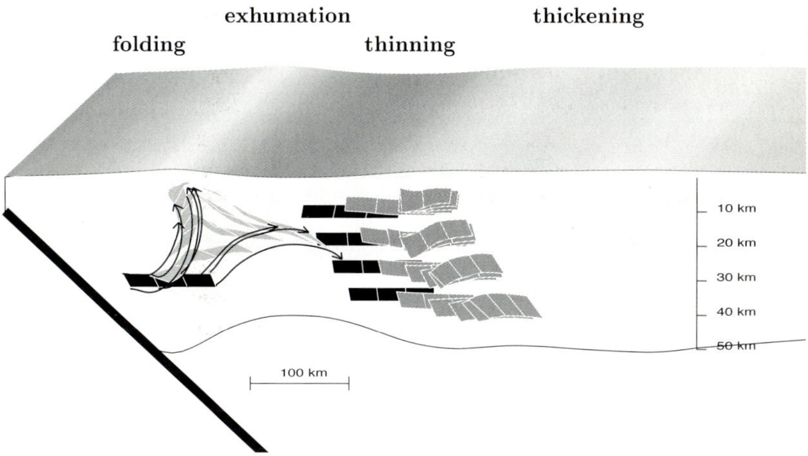

Fig. 2. Basic deformation pattern as a long-term result of the tilted shear (black boundary) mass redistribution on the surface and isostatic equilibration. Stages of internal mass movement are indicated (black-origin, grey-deformed). The original crust of 40 km thickness undergoes thickening in the front and the far field with a thinned section in between.

variation at the sheared boundary. A more detailed resolution of the internal mass flow is attained by a systematic representation of the movement of mass elements within a comparable period of time, Figures 3 and 4. Both figures reveal the corresponding pattern of mass movement for a change of the tilted shear from a dip angle of 15 to 25 degrees while the rate of surface erosion is lower in the models of Figure 3 $\left(D = 10^2\,\mathrm{m}^2/\mathrm{a}\right)$ compared to Figure 4 $\left(D = 10^3\,\mathrm{m}^2/\mathrm{a}\right)$.

A first view of the modified experiments confirms the issue about the evolution of a common structural pattern within the frame of parameter variation. However the particular attributes vary quantitatively: a low diffusion constant corresponds to a weak erosion rate below a millimeter per year. Obviously in the model experiments in Figure 3 the tectonic erosion through interplate shear dominates the long–term mass transport pattern compared to the influence of surface erosion. Primarily, this leads to thickening of the crust from 40 km to a maximum thickness of 70 km. The loop of induced mass transport goes deep. Thus the pressure and temperature regime which affects the moving material is rather wide. In detail we calculate pressure signatures on the surface of 11 kbar for the matter exposed within the eroded window for the 15 degree example, see Figure 3. According to the indicated mass flow the zone of upstream which forms the exhumation zone is fairly broad. It is evidently controlled by the dip angle of basal shear as the two examples indicate. For a steeper dip of 25 degrees the exposure of matter at the surface covers a smeared–out extended zone with a low level of pressure modification at about 5 kbar. This is indicated by the

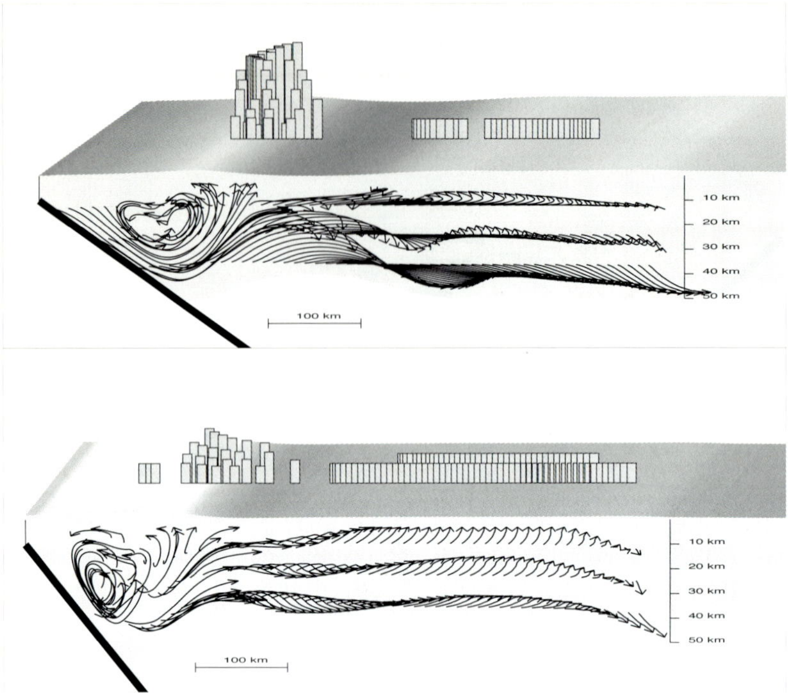

Fig. 3. Long-term flow pattern for low surface erosion $(D = 10^2\ \mathrm{m}^2/\mathrm{a})$ compared to tectonic erosion. Dip angle of the sheared boundary 15 degree (above) and 25 degree (below). Columns along the surface indicate the maximum pressure signature of exposed mass elements along their path of movement.

columns at the model surface which represent the pressure history as a function of the transport loop. Thus, whenever tectonic erosion dominates surface erosion effects, we see strong and deep–seated material transport which leads to the exhumation of high pressure metamorphic rocks only in the case of shallow subduction. From the model specification one expects only minor sediment supply in this case. They follow the major downflow of matter and thus form a broad zone of deposition at the foreland on top of the tip of the wedge. The hinterland basin reaches only 50 % of the width of the foreland basin.

The competition between tectonic and surface erosion produces alternative scenarios whenever the climatic erosion prevails. This is the case in the models of Figure 4, where the diffusion constant is $D = 10^3\ \mathrm{m}^2/\mathrm{a}$. The frontal mass flow pattern reaches a much smaller dimension in both directions. This causes less thickening of the crustal layer (up to 65 km). Instead of the broad ascent of mass

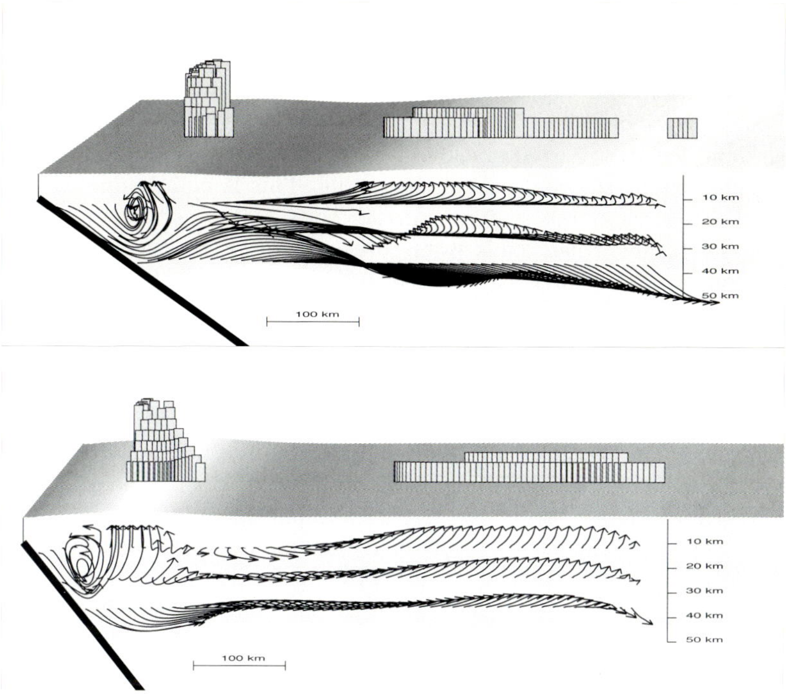

Fig. 4. Long-term flow pattern for high surface erosion ($D = 10^3\,\mathrm{m}^2/\mathrm{a}$) compared to tectonic erosion. Dip angle of the sheared boundary 15 degree (above) and 25 degree (below). See caption of Figure 3.

and its lateral bending in the uppermost zone, here the uprise is more straight and leads to a narrow window of exhumation (maximum pressure signature 9 kbar). The return flow is controlled by surface erosion and redistribution of matter. Because of the minor dimension of the overturning flow in the front zone, the forces produce an extremely wide thinning structure in the hinterland of the topographic uplift. This depression accumulates the large amount of sediments which find no way back into the recycling at the subduction zone. The dynamic uplift of the plateau on the right of the wedge is more extended laterally for low erosion, Figure 3, because of the sediment recycling and internal mass flow. In every case an increase of the dip of the shear boundary goes together with a distinct upthrust mode of deformation at the plateau far from the subduction zone. Obviously the development of the entire set of structural features is highly interdependent and linked to each other.

5 Discussion and conclusions

The benefit of the presented model approach is clearly the nonlinear coupling of tectonic and climatic erosion during dynamic mass flow on a geological time scale. For this purpose many details had to be parameterised and simplified in order to attain consistency of scales. The calculated patterns of mass movement exhibit a remarkable sensitivity to alteration of the load system, structural parameters and to some extent the rheology of the model. This is mainly due to the fully developed feedback in the system. Furthermore, this sensitivity concerns the entire model frame which leads even to implications for the far field of the sheared boundary. In particular we like to conclude that the dynamic model approach corresponds on the chosen scale with a large number of structures and tectonic features in nature.

- Shear induced by subduction has the ability to drive long-ranging deformation processes with the support of climatic erosion.
- Surface erosion on the other hand controls the internal pattern of dynamic mass movement: the formation of the orogenic root, and the existence, position, width and pressure level of the exhumation window.
- Surface erosion controls not only about the amount of released sediment but also whether they become recycled via the foreland or permanently stored in the hinterland. Of course, those interactions are only possible in the presence of the shear induced mass flow.
- On the scale of the entire model dimension, the extension zone of crustal thinning between the orogenic swell and the outer plateau can be considered as an intramontane basin. Its size and existence is again proportional to the intensity of surface erosion.
- Although the tilt of the shear boundary was kept constant in our models, an increasing dip angle supports deeper flow and the lateral extent of the thickened crust. The latter aspect is also amplified by an increasing strength of the subcrustal mantle structure.
- Finally we might speculate on the potential of faulting on the base of our continuum approximation. This might be associated with divergent flow orientation like in the upper zone of exhumation. There is a good chance for major fault formation. The outer plateau on the other hand is in a distinct condition of thrusting.

Acknowledgement

We would like to thank the Deutsche Forschungsgemeinschaft for financial support of the first author.

References

Beaumont, C., Fullsack, P., Hamilton, J., 1994. Styles of crustal deformation in compressional orogens caused by the underlying lithosphere. Tectonophysics, 232, 119–132.

Byerlee, J.D., 1978. Friction of rocks. Pure appl. Geophys., 116, 615–626.

Chester, F., 1995. A rheologic model for the wet crust applied to strike–slip faults. Journal of Geophysical Research, 100, 13033–13044.

Dahlen, F., 1984. Noncohesive Critical coulomb wedges: An exact solution, Journal of Geophysical Research, 89, 10125–10133.

Davis, D., Suppe, J., Dahlen, F., 1984. Mechanics of fold–and–thrust belts and accretionary wedges. Cohesive coulomb theory. Journal of Geophysical Research, 89, 10087–10101.

Gottschalk, G., 1997. A numerical model on the long-term dynamics of the continental lithosphere at convergent plate boundaries. Doctoral dissertation, University Bonn, 72p (in german).

Gutscher, M.-A., Kukowski, N., Malavieille, J & Lallemand, SE., 1998. Material transfer accretionary wedges from analysis of a systematic series of analog experiments. Journal of Structural Geology, 20, 4, 407–416.

Lallemand, SE., Schnürle, P. & Malavieille, J., 1994. Coulomb theory applied to accretionary and non-accretionary wedges: possible causes for tectonic erosion and/or frontal accretion. Journal of Geophysical Research, 99, 12033–12055.

Tichelaar, B.W., & Ruff, L., 1993. Depth of seismic coupling along subduction zones. Journal of Geophysical Research, B 98. 2017–2037.

Approximate theory of substratum creep and associated overburden deformation in salt basins and deltas

Florian K. Lehner

Geodynamics – Physics of the Lithosphere, University of Bonn, Germany
Tel. 0228 – 737429-30 – Fax: 0228 – 732508 – e-mail: lehner@geo.uni-bonn.de

Abstract. Lubricating squeeze flow of a ductile substratum under varying overburden is a characteristic element of the tectonics of salt basins and many deltas. Large-scale mass movements of salt or overpressured shales can occur in this manner, accompanied by deformation of a sedimentary overburden. The theory outlined in this paper deals with large-scale extrusive flow in mobile substrata that is driven by differential loads and buoyancy forces. The theory assumes slowly varying overburden and substratum thicknesses. It treats the salt (or shale) substratum as a viscous 'lubricating layer' and the overburden as a dead load that offers no significant resistance to shear along vertical planes, but will support horizontal stresses within the limits of 'active' and 'passive' Coulomb plastic states. Assuming the sea floor (sedimentation boundary) and basement elevations given, the theory yields a single differential equation in the layer thickness h of the salt (or shale) layer. When buoyancy effects are negligible, squeeze flow in a substratum of varying thickness will propagate substratum *isopachs* as 'kinematic waves' with a speed proportional to $(\tan\alpha)^n h^{n+1}$, where $\tan\alpha$ is the overburden slope and n the power law exponent governing the creep response of the substratum. Buoyancy introduces nonlinear unstable ('backward') diffusion effects leading to localized flow reversal and suggesting a mechanism for the generation of 'pinch-and-swell' structures; it also enters as the principal driving force for the large-scale updip extrusion of salt in areas affected by differential subsidence. The theory also sheds light on a frequently observed form of gravity-induced slope failure, which is characterized by the simultaneous occurrence of extensional and compressive faulting in distinct sections of the slope.

1 Introduction

The last two decades have brought many advances in the understanding of a particular tectonic and structural style that is characteristic for rheologically stratified systems in which a brittle *overburden* layer is underlain by a highly mobile, ductile *substratum*. The classical example is that of salt basins, but synsedimentary structures characteristic for 'salt tectonics' (Nettleton, 1955; Trusheim, 1960; Verrier & Castello Branco, 1972; Worrall & Snelson, 1989; Jackson &

Talbot, 1992; Jackson *et al.*, 1994) are also typical for major deltas, where they have been linked to differential loading of a mobile clay substratum (Bornhauser, 1958; Merki, 1972; Evamy *et al.*, 1978). Although it has long been recognized that substratum mobility must hold the key to synsedimentary structures such as turtle-back anticlines, pinch-and-swell structures or growth faults, little progress has been made in relating these phenomena quantitatively to substratum mobility. It therefore seemed appropriate to include in the present volume an early attempt in this direction (Lehner, 1973), published so far only in abstract form by the author (Lehner, 1977) and discussed briefly by (Mandl, 1988) and by (Last, 1988). This theory represents an attempt to quantify the squeeze flow induced in a mobile substratum by differential loads and to explore the structural consequences of such a flow for the overburden. Its first task has been the derivation of a partial differential equation describing the evolution of the substratum thickness in space and time; this is discussed in Section 2. Section 3 covers fundamental kinematic and mechanical aspects of substratum deformation and associated overburden failure by way of examples, exploring the effects of layer thickness on the mobility of the substratum as well as the role played by the salt-floor topography and by buoyancy in the early evolution of structures. The theory remains restricted to gentle deformations of the overburden/substratum boundary and—in an inessential manner—to negligible shearing resistance of the overburden along vertical planes. Possible further developments are summarized in a concluding discussion that also points to the use of the theory in the design of scaled physical analogue experiments.

2 Approximate theory of lubrication squeeze flow of a ductile substratum under a varying overburden

Consider the three-layer sequence shown in Figure 1, comprising a *basement*, *substratum*, and *overburden* and lying partly or wholly below sea level. For simplicity, the deformation of these strata is assumed to occur in the x, z coordinate plane, x being taken horizontal and z pointing vertically upwards. The main theoretical findings are readily generalized to three dimensions, as will be shown at appropriate places in the sequel. The layer boundaries are described by the elevations $z = h_k(x,t)$, $(k = 1, 2, 3)$, measured from a fixed datum level. The thicknesses of the overburden and substratum are given by the functions $H(x,t) = h_3(x,t) - h_2(x,t)$ and $h(x,t) = h_2(x,t) - h_1(x,t)$, respectively, while the water depth is denoted by $D(x,t)$ (cf. Fig. 1). Assuming that these functions possess the required continuity and smoothness properties, it is further stipulated that the substratum thickness h varies slowly with x and that it remains small everywhere compared to its longitudinal extent. The mobile substratum is treated as incompressible Newtonian viscous fluid, but a generalization of certain results, allowing for power-law creep behaviour, is given later in the paper.

This configuration of layers, with an overburden of variable thickness resting on a viscous substratum, may be considered typical for many salt basins and deltas, albeit as an idealization. It is also the configuration chosen in many

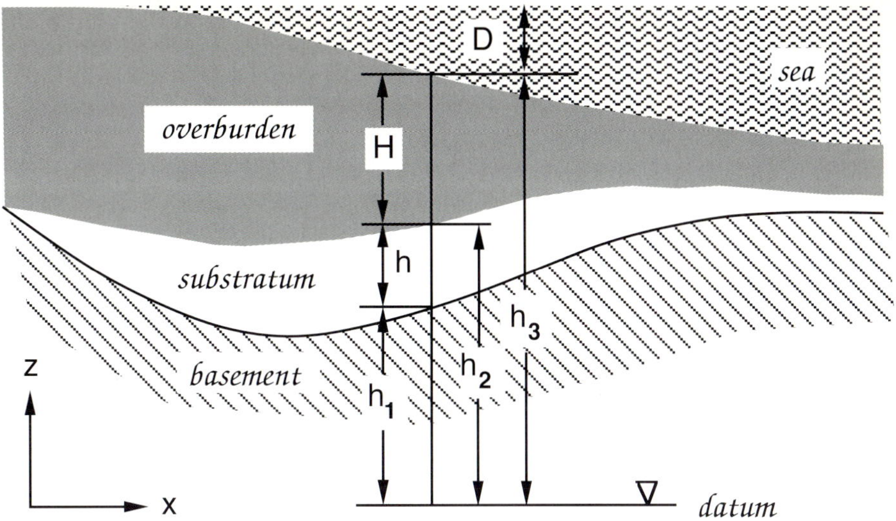

Fig. 1. Geometry of three-layer sequence

experimental studies. It must of course be seen as a system out of static equilibrium. In particular, one should expect the viscous substratum to respond to the uneven loading with some kind of *squeeze flow*.

In addressing this problem, the foregoing assumptions make it possible to invoke an approximate theory of lubrication squeeze flow (Langlois, 1964). It is shown there that horizontal velocity component v_x of the anticipated squeeze flow in the substratum is approximately be given by

$$v_x = U_2 \frac{z - h_1}{h} + U_1 \frac{h_2 - z}{h} - \frac{1}{2\mu} \frac{\partial p}{\partial x} (z - h_1)(h_2 - z), \tag{1}$$

where $p(x, z, t)$ is the pressure in the fluid-like substratum, μ is its viscosity, and U_1 and U_2 are the translation velocities parallel to x of the basement and overburden at the substratum boundaries. Assuming a no-slip condition to apply along these interfaces[1], the first two terms in this expression represent a (Couette-flow) contribution to v_x resulting from the shear imparted on the substratum by the surrounding layers, while the third term represents the component of pressure-driven (Poiseuille-) flow.

A further result of lubrication theory is that the pressure distribution in any cross-section $x =$ const. remains approximately hydrostatic, so that

$$\partial p / \partial z + \rho_s g = 0, \tag{2}$$

[1] This condition can clearly become inappropriate in situations where the overburden/substratum boundary acts as a weak detachement surface.

where ρ_s is the constant density of the material in the substratum and g the acceleration of gravity. An integration yields the vertical distribution of the pressure in the substratum

$$p(x,z,t) = p[x,h_2(x,t),t] + \rho_s g[h_2(x,t) - z]. \tag{3}$$

Note that the pressure at the overburden/substratum boundary, which enters into this expression, is still unknown. Since $\partial p/\partial x$ does not depend on z, equation (1) immediately yields the total flow rate

$$q(x,t) = \int_{h_1(x,t)}^{h_2(x,t)} v_x \, dz = \frac{1}{2}h(U_1 + U_2) - \frac{h^3}{12\mu}\frac{\partial p}{\partial x}. \tag{4}$$

This shows once more that the total 'lubrication flow' in the substratum is composed of a Couette-flow and a Poiseuille-flow component.

The assumption of incompressibility demands that rate of change of the total flow rate q with x be balanced exactly by the rate of change in the layer thickness h with time, a condition expressed by the integrated continuity equation

$$\frac{\partial q}{\partial x} + \frac{\partial h}{\partial t} = 0. \tag{5}$$

Substitution of expression (4) for q then yields "Reynolds' equation"

$$\frac{\partial}{\partial x}\left[\frac{1}{2}h(U_1 + U_2) - \frac{h^3}{12\mu}\frac{\partial p}{\partial x}\right] + \frac{\partial h}{\partial t} = 0, \tag{6}$$

which is a fundamental result in lubrication theory (Langlois, 1964).

In the typical applications of lubrication theory, equation (6) serves to determine the pressure distribution for a given lubrication film thickness $h(x,t)$ and for given boundary velocities U_1, U_2. By contrast, the tectonic problems considered in this paper call for an interpretation of (6), in which the channel width $h(x,t)$ becomes the unknown dependent variable, while the pressure gradient is either independently known or expressible in terms of $h(x,t)$ and its gradient (Lehner, 1977). This simplest class of problems corresponds to the situation in which the pressure at the overburden/substratum boundary may be equated with sufficient accuracy to the weight per unit area imposed by the local solid and fluid overburden column, so that

$$p[x,h_2(x,t),t] = \rho_o g[h_3(x,t) - h_2(x,t)] + \rho_w g D(x,t), \tag{7}$$

where ρ_o denotes the bulk density of the overburden and ρ_w the density of sea water. One can write this in terms of a 'reduced density' $\rho_o' = \rho_o - \rho_w$ in the form

$$p[x,h_2(x,t),t] = \rho_o' g h_3(x,t) - \rho_o g[h(x,t) + h_1(x,t)] + \rho_w g[h_3(x,t) + D(x,t)]. \tag{8}$$

The pressure distribution (3) in the substratum now becomes

$$p(x,z,t) = \rho_o' g h_3(x,t) + \Delta\rho g[h(x,t) + h_1(x,t)] + \rho_w g[h_3(x,t) + D(x,t)] - \rho_s g z, \tag{9}$$

where $\Delta\rho = \rho_s - \rho_o$ throughout this paper.

At locations where the overburden top lies below sea level, the sum $h_3(x,t) + D(x,t)$ simply defines the sea level in coordinate system of Figure 1 and therefore does not depend on x. Assuming the land surface behind the shore line to remain essentially at the sea level[2], it follows that $h_3(x,t) + D(x,t) = \text{const.}$ everywhere. Differentiation of the above expression for the pressure and substitution of the result in (4) therefore yields the flow rate

$$q = \frac{1}{2}h(U_1 + U_2) - \frac{h^3}{12\mu}\left[\rho_o' g\frac{\partial h_3}{\partial x} + \Delta\rho g\left(\frac{\partial h}{\partial x} + \frac{\partial h_1}{\partial x}\right)\right] \tag{10}$$

and when this is used in (6), one obtains the following differential equation for the substratum thickness

$$\frac{\partial}{\partial x}\left\{\frac{1}{2}h(U_1 + U_2) - \frac{h^3}{12\mu}\left[\rho_o' g\frac{\partial h_3}{\partial x} + \Delta\rho g\left(\frac{\partial h}{\partial x} + \frac{\partial h_1}{\partial x}\right)\right]\right\} + \frac{\partial h}{\partial t} = 0. \tag{11}$$

It is clear that this result may be generalized immediately so as to describe lubrication squeeze flows in three dimensions, i.e., by simply adding a term involving the boundary velocities and spatial derivatives in the second horizontal, or y direction.

The following section deals with a number of special cases in which equation (11) can either be solved analytically or else can directly yield insight into the behaviour of a particular geological system in time. The principal limitations of (11) remain those of lubrication theory and of the assumed pressure distribution (9), which essentially implies a negligible shear resistance of the overburden. These will also be discussed in the following.

3 Mechanisms and structures associated with lubrication squeeze flow in ductile substrata

3.1 Kinematic waves

Let us begin with a problem (Lehner, 1973; Last, 1988) in which the differential load imposed by the overburden outweighs the effects of a possible density contrast between overburden and substratum, $\Delta\rho$ being therefore equated to zero in (11). The basement is assumed to remain fixed in its initial configuration, while the overburden is horizontally confined ($U_1 = U_2 = 0$) and allowed to deform

[2] This condition, which is appropriate for the applications contemplated in the present context, may obviously be relaxed, where necessary.

by nonhomogeneous shear along vertical planes only. In these circumstances, the expression for total squeeze flow rate becomes

$$q = (\rho'_o g \tan\alpha / 12\mu) h^3, \tag{12}$$

with $\tan\alpha = -\partial h_3/\partial x$, α being the instantaneous local slope angle of the overburden.

Squeeze flow under a constant differential load In order to develop an appreciation of certain basic phenomena, it will suffice to consider a constant slope angle α. The flow rate is then an explicit function $q(h)$ of h only and (12) and (5) yield the following simplified form of equation (11)

$$c(h)\frac{\partial h}{\partial x} + \frac{\partial h}{\partial t} = 0, \tag{13}$$

where

$$c(h) = dq(h)/dh = (\rho'_o g \tan\alpha / 4\mu) h^2. \tag{14}$$

Equation (13) is a quasilinear first-order partial differential equation and as such is of a type that arises in many different branches of physics and engineering, governing the propagation of flood waves in rivers, the flow of glaciers, the chromatography of solutes in solid beds, the transport of immiscible liquids through porous media, the flow of traffic on crowded roads, etc. (A standard reference is the book by Whitham (1974).) Associated with equation (13) is the notion of a *kinematic wave* and the equation itself—which typically arises from the combination of the conservation law (5) and a functional relation $q = q(h)$—has come to be known as *kinematic wave equation.* These terms were introduced by Lighthill & Whitham (1955) in order to distinguish wave motions of a purely kinematic origin (viz. in a continuity equation) from the classical wave motions encountered in dynamical systems that depend on Newton's second law of motion. Note that the quantity $c(h)$ not only has the dimensions of a velocity, but is also readily interpreted as the speed of propagation or *kinematic wave velocity* of positions $x(t)$ of constant layer thickness h. Indeed, since the condition

$$\frac{dh}{dt} = \frac{dx}{dt}\frac{\partial h}{\partial x} + \frac{\partial h}{\partial t} = 0 \tag{15}$$

must be satisfied continually in the propagation of constant values of h, a comparison of (13) and (15) shows that the set of *characteristic curves* $x(t)$ in the x,t-plane along which constant values of h are propagated are determined by the characteristic equations (Whitham, 1974)

$$dx/dt = c(h), \tag{16}$$

$$dh/dt = 0. \tag{17}$$

It is interesting to compare the wave velocity $c(h)$ with the mean particle velocity $q(h)/h$ for a given cross-sectional width h. From (12) and (14) it is seen that the kinematic wave velocity exceeds the mean particle velocity by a factor three!

A solution, by the *method of characteristics*, of the initial value problem for (13) with

$$h(x, t = 0) = h_0(x), \tag{18}$$

is now constructed from the two independent solutions

$$x - c(h)t = \xi, \tag{19}$$
$$h = \text{const.} \tag{20}$$

of the coupled pair of ordinary differential equations (16),(17). Here ξ is an integration parameter with the significance of an initial point on a characteristic curve. Since h remains constant along a characteristic curve, it must remain equal to its initial value $h_0(\xi)$. The solution of the initial value problem is therefore obtained in the following parametric form

$$x = \xi + c(h)t, \tag{21}$$
$$h = h_0(\xi). \tag{22}$$

Alternatively, the parameter ξ may be eliminated so as to render the solution of the initial value problem in the implicit form

$$h = h_0(x - c(h)t). \tag{23}$$

This shows clearly that the positions x_1 and x_2 occupied by a given layer thickness h at two different times t_1 and t_2 are related by $x_2 - x_1 = c(h)(t_2 - t_1)$ and thereby establishes again the significance of the propagation speed $c(h)$.

Before displaying the solution (21),(22) graphically, it will be convenient to bring it into a dimensionless form by introducing the following normalizations

$$\bar{h} = h/l, \qquad \bar{x} = x/l, \qquad \bar{\xi} = \xi/l, \qquad \bar{t} = t/\tau_s, \qquad \bar{c} = c\tau_s/l. \tag{24}$$

Here l is a characteristic length of the system and τ_s is a characteristic time that is suitably chosen as

$$\tau_s = 4\mu/(\rho_o' gl \tan\alpha). \tag{25}$$

The differential equation (13) now assumes the form

$$\bar{c}(\bar{h})\frac{\partial \bar{h}}{\partial \bar{x}} + \frac{\partial \bar{h}}{\partial \bar{t}} = 0, \tag{26}$$

while (21),(22) become

$$\bar{x} = \bar{\xi} + \bar{c}(\bar{h})\bar{t}, \tag{27}$$
$$\bar{h} = \bar{h}_0(\bar{\xi}), \tag{28}$$

the dimensionless wave velocity being given by $\bar{c}(\bar{h}) = \bar{h}^2$.

Figure 2 illustrates the graphical construction, from (27) and (28), of the kinematic wave motion produced by a differential load in a substratum of a given initial thickness distribution $\bar{h}_0(\bar{\xi})$. One first draws a number of characteristic *base lines* (27), covering the relevant range of initial values $\bar{\xi}$ and subsequently lays off vertically the constant layer thickness $\bar{h} = \bar{h}_0(\bar{\xi})$ propagated along each base line at successive times $\bar{t}_1, \bar{t}_2, ...$, measuring $\bar{h}$ upwards from the (fixed) basement level. In this manner, the solution surface of equation (26) is represented directly by the shape of the overburden/substratum interface $\bar{z} = \bar{h}_2(\bar{x}, \bar{t})$ as seen in cross-section at successive times.

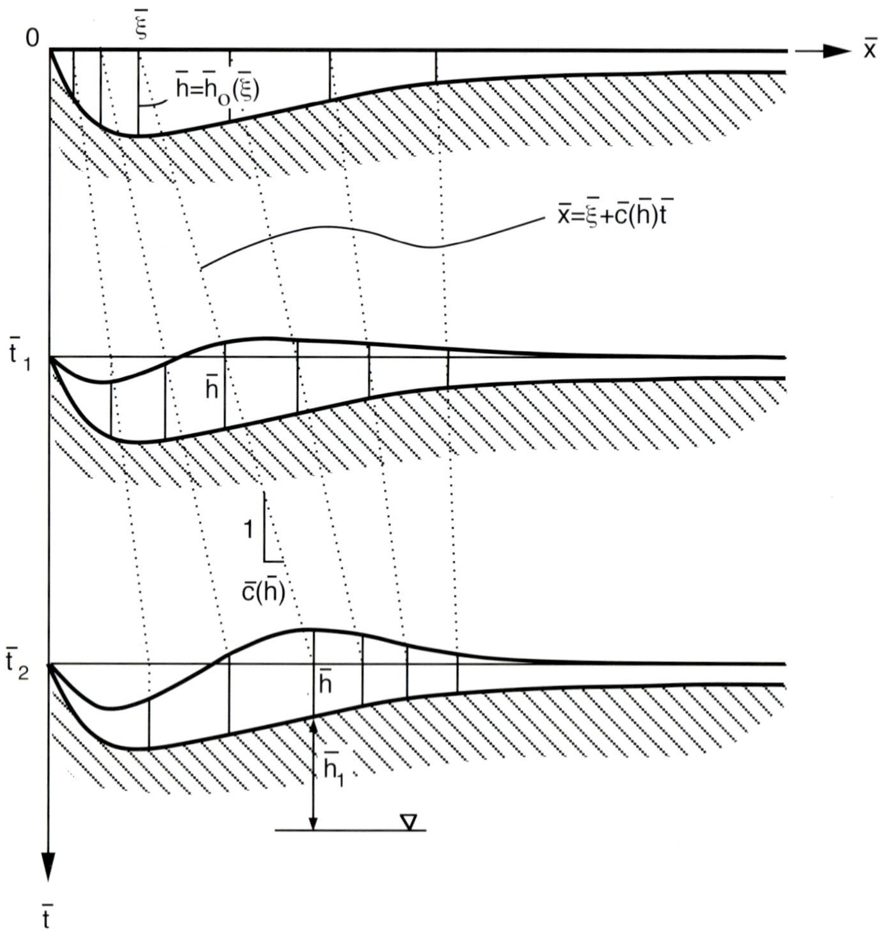

Fig. 2. Graphical solution of kinematic wave equation for given initial data

Figure 2 also highlights the following characteristic features of kinematic wave motion. First of all, the distinction between wave and particle velocities: While each layer thickness $\bar{h}$ is propagated along the basement relief at its own wave speed $\bar{c}(\bar{h})$, material points that lie on the overburden/substratum boundary rise and fall without undergoing any horizontal displacement. The initially thickest section of the substratum is propagated along the basement relief in the form of a travelling anticline. The anticline steepens, because the wave speed increases with $\bar{h}$ so that the larger layer thicknesses tend to overtake the smaller ones lying ahead. Mathematically, the solution for the wave shape thus becomes multivalued at a certain point (Fig. 3). For $dc/dh > 0$, one can easily see therefore that the appearance of a multivalued solution is due to the fact that $dh_0/d\xi < 0$ for some ξ at least. Multivaluedness must however be excluded on physical grounds for most processes that are governed by quasilinear partial differential equations. Flood waves tend to break and concentration gradients can steepen only to the point where a concentration jump appears, etc., and additional arguments must therefore be brought in in the form of 'shock conditions' that allow regions of multivaluedness to be replaced by hydraulic, concentration, or other kinds of 'jumps' or 'shocks' that travel at their own speed (Whitham, 1974). In the present context discontinuous solutions of this type must nevertheless be ruled out, primarily because the proposed theory itself remains restricted in its validity by the underlying assumption of slowly varying layer thicknesses[3].

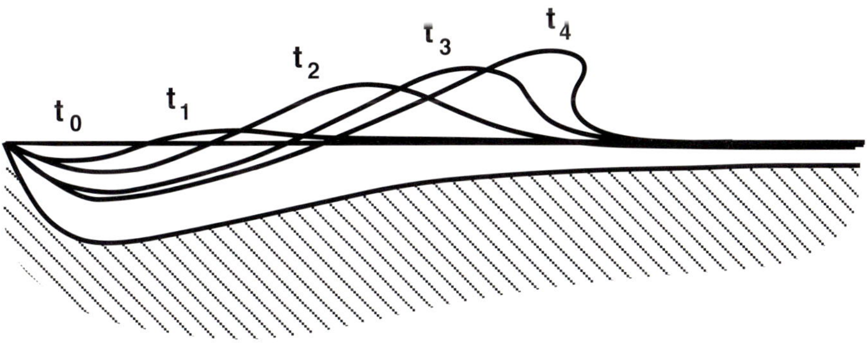

Fig. 3. Multivalued solution of kinematic wave equation

[3] An imaginative reader might be tempted to interpret multivalued solutions physically in terms of 'nappes' that override a layer of 'mother salt'. Although the present theory must not be 'pushed that far' for the above reason, the fact that it predicts an evolution towards such multivalued solutions from within its range of validity remains suggestive and can form an incentive for further studies.

Another feature of the solution of the initial value problem for equation (26) is its dependence at any point $(\bar{x}, \bar{t})$ on only the 'upstream' part of the initial data. This is quite apparent from Figure 2. However, the principle point illustrated by the above example remains the phenomenon of propagation of substratum *isopachs* across a given basement relief and the structural imprint (e.g., in the form of anticlinal ridges or swells) that is thereby created in the overburden.

Time-dependent differential load Let us now slightly generalize the above problem by admitting a time-dependent rather than constant overburden slope, i.e., let us assume that the slope steepens monotonically in time from zero to some fixed maximum value $\tan\alpha$, for example according to a relation of the form

$$-\partial h_3/\partial x = (1 - e^{-t^2/\tau_o^2}) \tan\alpha. \tag{29}$$

As can be seen, this represents the more realistic situation where the system is initially in equilibrium and where the slope angle increases at a rate that is characterized by the time constant τ_o. It is possible in this case, to preserve equation (26) in the same nondimensional form by a suitable change in the time scale. For example, given (29), one replaces t by a function $t'(t)$ such that $dt'/dt = 1 - \exp(-t^2/\tau_o^2)$, i.e.,

$$t'(t) = t - (\sqrt{\pi}/2)\tau_o \,\mathrm{erf}\,(t/\tau_o), \qquad \text{where} \quad \mathrm{erf}\,x = \frac{2}{\sqrt{\pi}} \int_0^x e^{-\lambda^2}\, d\lambda \tag{30}$$

is the error function. Also, dimensionless time is redefined as $\bar{t} = t'/\tau_s$, where τ_s is the same as in (25), with the understanding that $\tan\alpha$ now represents the final, maximum slope.

Given identical initial conditions, the dimensionless form of the solution for a time-dependent slope will thus agree completely with that for a fixed slope. The difference betwen the solutions becomes apparent only if real time t is plotted along the time axis in the example of Figure 2. In other words, the substratum passes through the same sequence of shapes in the course of its deformation, but the process is delayed if the overburden steepens gradually. Whether or not the delay is significant will depend on the ratio τ_s/τ_o of the characteristic times. Indeed, when writing the solution (27),(28) in terms of the original time variable t

$$\bar{x} = \bar{\xi} + \bar{h}_0^2(\bar{\xi})\{t/\tau_s - (\sqrt{\pi}/2)(\tau_s/\tau_o)^{-1}\mathrm{erf}\,[(t/\tau_s)(\tau_s/\tau_o)]\}, \tag{31}$$

the role of the scaling parameter τ_s/τ_o is immediately apparent. Thus, in the extreme cases $\tau_s/\tau_o \ll 1$ and $\tau_s/\tau_o \gg 1$ the solution approaches the limits $\bar{x} \rightarrow \bar{\xi}$ and $\bar{x} \rightarrow \bar{\xi} + \bar{h}_0^2(\bar{\xi})t/\tau_s$, corresponding to an infinitely slow and instantaneous build-up of the differential load, respectively.

Solution for power-law creep behaviour Rock salt is known to exhibit power-creep (Carter *et al.*, 1993) and similar behaviour has been assumed occasionally for shales, although the rheology of overpressured shales is likely to be much more complex. However, a critical evaluation of currently advocated rheological models for shales lies beyond the scope of this paper. The following results for power-law creep may nevertheless point in the right direction, if one wishes to take a step beyond linear viscous behaviour.

Consider therefore the nonlinear constitutive law (in Cartesian component notation) for the stress σ_{ij} as a function of the rate of deformation $D_{ij} = 1/2(\partial v_i/\partial x_j + \partial v_j/\partial x_i)$:

$$\sigma_{ij} = -p\,\delta_{ij} + 2\sigma_0 t_0^{\frac{1}{n}} \mathrm{II}_D^{\frac{1-n}{2n}} D_{ij}, \quad n \geq 1. \tag{32}$$

Here σ_0 and t_0 represent a reference stress and time. The power-law exponent n of a Newtonian viscous fluid equals 1, while for rock salt values of n between 3.4 and 5.3 have been proposed (Carter *et al.*, 1993) for strain rates between 10^{-5} and $10^{-9}\,\mathrm{s}^{-1}$. The quantity II_D denotes the second invariant of the deformation rate. For the plane, incompressible lubrication flows under consideration $1/2(\partial v_x/\partial z)$ is the only significant component of the deformation rate (in the coordinate system of Figure 1). The general expression $\mathrm{II}_D = \frac{1}{2} D_{ij} D_{ij}$ therefore reduces to $\mathrm{II}_D = 1/4(\partial v_x/\partial z)^2$. The relevant shear stress component for lubrication flow in the $x-$direction is therefore given by

$$\sigma_{xz} = \sigma_0 t_0^{\frac{1}{n}} \left| \frac{\partial v_x}{\partial z} \right|^{\frac{1-n}{n}} \frac{\partial v_x}{\partial z}. \tag{33}$$

The geometrical and kinematic assumptions on which lubrication theory is founded also imply the approximation $\partial \sigma_{xx}/\partial x \approx -\partial p/\partial x$, which together with (33) yields the equilibrium equation

$$\frac{\partial \sigma_{xz}}{\partial z} + \frac{\partial \sigma_{xx}}{\partial x} = \sigma_0 t_0^{\frac{1}{n}} \frac{\partial}{\partial z} \left(\left| \frac{\partial v_x}{\partial z} \right|^{\frac{1-n}{n}} \frac{\partial v_x}{\partial z} \right) - \frac{\partial p}{\partial x} = 0. \tag{34}$$

Since $U_1 = U_2 = 0$, the flow must be symmetric about the center of the layer so that an integration of (34) over the layer thickness gives

$$\tau_{xz}(h_1) = -\tau_{xz}(h_2) = -\frac{h}{2}\frac{\partial p}{\partial x}. \tag{35}$$

The first integral of (34) gives

$$\sigma_0 t_0^{\frac{1}{n}} \left| \frac{\partial v_x}{\partial z} \right|^{\frac{1-n}{n}} \frac{\partial v_x}{\partial z} = -\frac{\partial p}{\partial x} \left(z - \frac{h_1 + h_2}{2} \right) \tag{36}$$

and integrating once more,

$$v_x = \frac{-1}{(n+1)\sigma_0^n t_0} \left| \frac{\partial p}{\partial x} \right|^{n-1} \frac{\partial p}{\partial x} \left[\left(\frac{h}{2} \right)^{n+1} - \left| z - \frac{h_1 + h_2}{2} \right|^{n-1} \left(z - \frac{h_1 + h_2}{2} \right)^2 \right]. \tag{37}$$

A final integration over the layer thickness yields the total flow rate

$$q = \frac{-1}{(n+2)2^{n+1}\sigma_0^n t_\circ} \left|\frac{\partial p}{\partial x}\right|^{n-1} \frac{\partial p}{\partial x} h^{n+2}. \tag{38}$$

Under the above assumptions of a uniform differential load and negligible density contrast between overburden and substratum the flow rate becomes

$$q(h) = [(\rho_o' g \tan\alpha)^n / ((n+2)2^{n+1}\sigma_0^n t_\circ)] h^{n+2} \tag{39}$$

and the kinematic wave velocity is therefore given by

$$c(h) = dq(h)/dh = [(\rho_o' g \tan\alpha)^n / (2^{n+1}\sigma_0^n t_\circ)] h^{n+1}. \tag{40}$$

The solution to the initial value problem for (26) may now be recovered in the dimensionless form (27),(28) by normalizing again as in (24), while replacing (25) by

$$\tau_s = (2^{n+1}\sigma_0^n t_\circ)/(\rho_o' g l \tan\alpha)^n. \tag{41}$$

wave velocity is thus given by $\bar{c}(\bar{h}) = c(h/l)\tau_s/l = \bar{h}^{n+1}$ in the present case. Moreover, the true wave velocity (40) is proportional to $(\tan\alpha)^n$. The ratio of the kinematic wave velocity to the mean particle velocity of the flow is now $c(h)h/q(h) = n+2$. This difference between linear viscous and power-law creep behaviour has important consequences. For example, the much stronger dependence of the kinematic wave velocity on layer thickness and driving force found for power-law creep behaviour may well explain why significant salt mobility often appears associated with a certain minimum layer thickness. If one takes the kinematic wave velocity as an appropriate measure of 'mobility', bearing in mind that it is the mobility of structural features such as moving layer boundaries that is typically observed (i.e., reconstructed), then the ratio $c^{(n)}/c^{(1)}$ of the kinematic wave velocities for $n > 1$ and $n = 1$ will be the quantity of interest. From the above results this ratio is given by

$$c^{(n)}/c^{(1)} = h^{n-1}/h_{\rm crit}^{n-1}, \tag{42}$$

provided the constant Newtonian viscosity μ is equated to $\sigma_0 t_0$. Here $h_{\rm crit} \equiv 2\sigma_0/(\rho_o' g \tan\alpha)$ plays precisely the role of a critical layer thickness: for a power-law fluid the mobility of squeeze-flow structures remains subdued relative to that of a Newtonian fluid for $h < h_{\rm crit}$, while the opposite is true for $h > h_{\rm crit}$.

The problems discussed so far are probably the simplest in a class, but as such may serve best to illustrate an important nonlinear wave phenomenon affecting mobile substrata under differential loads. The picture of *substratum isopachs* that are propagated with a given characteristic velocity across the basement relief may prove to be of considerable help in the interpretation of certain structures, especially those belonging to an early, pre-diapiric stage of salt tectonics. As Kupfer (1974) has observed in a comprehensive and thoughtful review some twenty-five years ago: "The evidence of early halokinesis is important and growing, and generally unexplained by density differences. Its importance cannot be underestimated, as these early movements determine the pattern of later deformations."

3.2 Squeeze flows driven or modified by buoyancy forces

It is true, of course, that negative density differences ($\Delta\rho < 0$) have been linked to the early stage of salt tectonics ever since Arrhenius' proposal in 1912 of a buoyancy-driven rise of salt domes (cf. Nettleton (1955) and Trusheim (1960) for reviews of the early history of the subject). It is interesting to note the important contribution made by early experimental studies, employing ductile as well as brittle model materials, to the evolution of these ideas. During a subsequent period, the study of salt tectonics came under the spell of theoretical work in fluid dynamics that appeared to hold the key—in the form of the classical RALEIGH-TAYLOR instability—to all salt tectonic phenomena. But a change in emphasis came again with a revival of the earlier experimental work and a renewed effort to model the difference in mechanical behaviour of ductile substrata and brittle overburden layers. Experimentally produced structures could now be compared with much improved seismic images of real structures. As a result of such studies, diapirism for example is now seen to be linked in an important way to overburden extension. The somewhat dogmatic heritage of a purely fluid-dynamical interpretation of salt diapirs has thus been replaced by a more refined picture[4] that distinguishes different modes of diapirism (Vendeville & Jackson, 1992; Jackson & Vendeville, 1994; Jackson *et al.*, 1994).

The present theory can shed more light on the role of a density contrast as a primary driving force or modifying factor of lubrication squeeze flows, which will affect also the initiation of diapirs. Going back to equation (11), the first thing to notice when $\Delta\rho \neq 0$ is the appearance of the nonlinear second-order term in h. Reynolds' equation thus turns into a diffusion equation with a nonlinear diffusivity given by $\Delta\rho g h^3/12\mu$. The kinematic wave motion associated with the

[4] In this picture, an initiation of diapirs by buckle folding has been ruled out, largely on the basis of experimental observations. The fact remains, however, that experiments employing sand as a model overburden are not fully scaled, while potentially conclusive theoretical work on realistic salt/overburden (fluid/solid) systems is still in a relatively early stage, having been concerned mostly with the problem of layer instability as affected by overburden strength, in-situ stress and layer thickness (Leroy & Triantafyllidis, 1996; Triantafyllidis & Leroy, 1997; Leroy & Triantafyllidis, this volume). Massin *et al.* (1996) have studied the nonlinear evolution of buckle folds in an elastoplastic overburden numerically up to the onset of faulting, when the layer was found to collapse in a chevron mode. This type of analysis clearly needs to be pursued further to account for the effects of erosion and sedimentation and in particular, to focus on shorter wavelength perturbations that could result in a mode of collapse facilitating the growth of piercement diapirs. It is clear that the results of such an analysis would be very difficult to reproduce experimentally, primarily because of the inherent difficulty of achieving a full scaling of the nonlinear elastoplastic response of the overburden. This concerns in particular the relative timing of buckle folds and thrust faults in a compressive regime. The partial or 'extended similitude' (Mandel, 1962), which is achieved when using sand as a brittle model material, may be well suited for extensional tectonic regimes, but may fail to reproduce the more subtle competition between folding and faulting in thin-skinned compression, where in fact it will be biased strongly towards faulting.

convective first-order terms will thus be modified by a diffusive decay of the layer thickness h, if $\Delta\rho > 0$[5], or an amplification, if $\Delta\rho < 0$, the latter case corresponding to a gravitationally unstable stratification. Note however that equation (11) is not well posed mathematically for $\Delta\rho < 0$.[6] The reason for this defect must be seen in the neglect of a higher order term representing an average shear resistance of the overburden. Although it would be possible to obtain such a term from one or the other approximate treatment of overburden resistance, such a development lies beyond the scope of this paper. It is therefore fortunate that one can make a number of interesting inferences directly from expression (10) for the total flow rate q, without having to solve equation (11).

For simplicity, let $U_1 = U_2 = 0$ in (10) and consider the condition

$$\tan\beta_{\text{crit}} \equiv \left(\frac{\partial h_2}{\partial x}\right)_{\text{crit}} = \left(\frac{\partial h}{\partial x} + \frac{\partial h_1}{\partial x}\right)_{\text{crit}} = -\frac{\rho_o'}{\Delta\rho}\frac{\partial h_3}{\partial x} \tag{43}$$

for the vanishing of q, assuming that $\Delta\rho \neq 0$. The condition thus demands that the slopes of the sedimentation boundary (or top overburden) and overburden/substratum interface satisfy a simple proportionality relation. If $\Delta\rho < 0$, both interfaces must dip in the same, they must dip in the opposite direction. For a buoyant substratum, this also implies a reversal of the sign of q, i.e., a reversal of the bulk flow at locations where the slope of the overburden/substratum boundary steepens beyond the above critical value. As illustrated in Figure 4, this must lead to a localized upwelling of material, accompanied by the development of rim synclines around the point of flow reversal (location A in Fig. 4b). A second generation anticlinal ridges may grow with the appearance of new starvation zones in locations such as B and C in Fig. 4c. The process may generate several generations of anticlinal ridges or swells in this way, being controlled not only by the overburden gradient but also in an essential manner by the basement topography.

---

[5] Bird (1991), Buck (1991), and Hopper & Buck (1996) have employed diffusion equations, corresponding to special cases of equation (11) for $\Delta\rho > 0$, to study the flow of lower crustal rocks under topographic highs and during thick-skinned lithospheric extension.

[6] This is to say that any smooth solution to the equation, if it exists at all for given initial data after some finite time, may differ arbitrarily from a second solution for only infinitesimally different initial data. In general, a solution for smooth initial data will tend to develop singular peaks after some finite time. This can easily be seen from the fact that in its simplest form, i.e., when $\partial h_3/\partial x = \partial h_1/\partial x = 0$ and $U_1 = U_2 = 0$, equation (11) reduces to the so-called 'backward heat equation' $\partial/\partial x(D(h)\partial h/\partial x) = -\partial h/\partial t$ which describes a diffusive process that is viewed backward in time. Thus, suppose one were first to run this process forward in time for given initial data involving a number of singular peaks. The smooth solution to this forward problem at some finite time t_1 could then be taken to represent the initial data for a backward problem in the time variable $t_1 - t$. The solution of the backward problem will thus become singular at $t_1 - t = t_1$, when it assumes the form of the initial data for the forward problem. In general, however, one would not only be unable to predict t_1 but also unable to solve the nonlinear backward equation numerically.

In view of what has been said about the non-wellposed character of equation (11), when $\Delta\rho < 0$, the structural evolution sketched in Figure 4 must be seen as a qualitatively expected rather than quantitatively predicted one. Apart from a certain vertical exaggeration, that should not be misinterpreted in terms of salt diapirism, the structures drawn are as smooth as one would expect of such early anticlinal and synclinal features in nature. The main message to be read from Figure 4 is that differential loading and buoyancy effects in combination with the topography of the salt floor are likely to determine the early structural pattern of the top salt and consequently must have a decisive influence on the entire salt tectonic history of a basin.

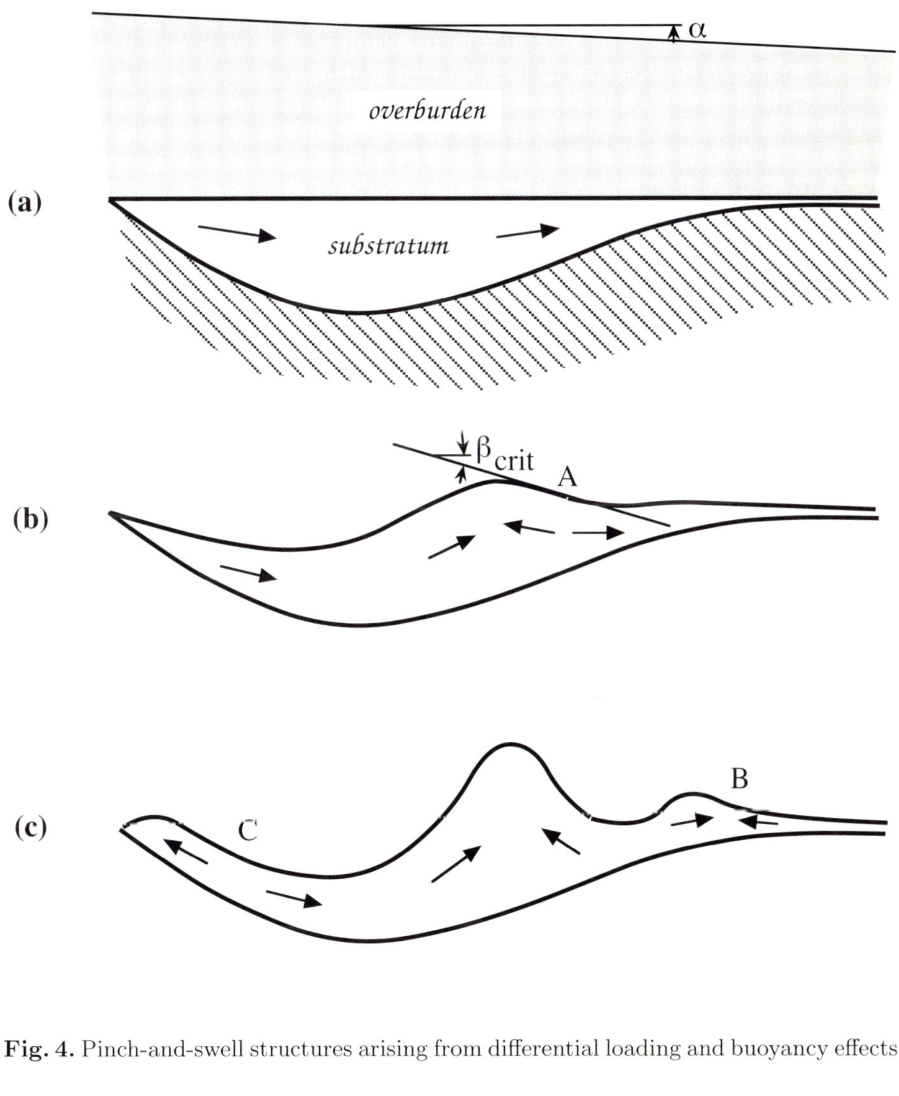

Fig. 4. Pinch-and-swell structures arising from differential loading and buoyancy effects

The theoretically expected pinch-and swell structures shown in Figure 4 are of a type that has been often been observed in mechanically similar geological settings. Bornhauser (1958) cited the early experimental work of Rettger (1935) in support of his gravity-flow concept of Gulf coast tectonics, according to which "Anticlinal and synclinal folding appears to have been initiated by the gravity flow of incompetent salt or shale beds down a sloping surface under sedimentary load"[7]. Morgan *et al.* (1968) explain the pinch-and-swell structures in mobile and buoyant overpressured shales at the mouth of the Mississippi River in terms of a prograding differential load. On a much smaller scale, the same type of phenomenon has been described in lagoonal sediments, where muds are forced into a series of ridges at the toe of an advancing sand dune which is cut by normal (growth) faults as it settles into the underlying mud (Brown, 1969).

Pinch-and-swell structures from which diapirs may develop by downbuilding are a major element of the picture of salt deformation in the Gulf of Mexico that has emerged in recent years, especially with the discovery of spreading, allochtonous salt sheets (Worrall & Snelson, 1989; Humphris, 1978,1979; West, 1989). The allochtonous nature of major salt 'tongues' that spread across relatively undisturbed (progressively) younger deep-sea sediments has also been discussed in some detail by Jackson & Talbot (1992). In this picture, differential loading and buoyancy effects are assumed to cooperate to produce the overall extrusive flow and the characteristic seaward progression of structures. Where the frontal portion of the spreading salt sheet is covered by only a thin veneer of sediments, the absence of large amplitude diapirs may be explained by insufficient sediment supply. The flow of salt changes from a lubrication squeeze flow to a gravitational spreading mode, in which the thin sediment cover offers no significant restraint. The flow resembles that of a large ice sheet. The present theory could be appropriately modified to account for this transition, making use of a well-developed body of theory (Nye, 1963; Paterson, 1969). It could also be based on (5) with a nonzero 'sink term' on the right-hand-side, representing the loss of mass from the salt sheet by dissolution. Such a theory may be worth pursuing in the specific context of the Gulf coast salt sheets,

Travelling salt ridges have also been discussed by Trusheim (1960), who observed that structural salt trends in Northern Germany run approximately parallel to strike of the dipping basement. Trusheim takes this as an indication of regional salt migration from the deepest part of the basin into border areas (such as the Mansfeld basin in Germany), where structural trends parallel to the rim of the basin are evident. The connection seen by Trusheim between regional salt migration and the configuration of the basin floor is readily explained by the present theory. Thus, for $U_1 = U_2 = 0$ but $\Delta\rho < 0$ expression (10) for the flow

[7] Worrall & Snelson (1989) appear to have read into Bornhauser's (unnecessary) assumption of a 'sloping surface' an endorsement of *gravity sliding*, a mechanism against which they argue convicingly in favour of *gravitational spreading* as discussed here. In fact, Bornhauser clearly distinguishes between his preferred hypothesis of "sedimentary creep" and that of a "slumping or gliding type of gravity flow which proceeds at a faster rate and moves along a glide plane".

rate clearly predicts an updip flow (i.e., flow in the direction of increasing h_1) as long as the condition

$$\frac{\partial h_1}{\partial x} > \frac{\partial h}{\partial x} - \frac{\rho_o'}{|\Delta\rho|}\frac{\partial h_3}{\partial x} \tag{44}$$

is fulfilled. As soon as it is violated at a given location, the total flow will be reversed at that location (for example, location C in Fig. 4), and this again may signal the beginning of a sequential development of pinch-and-swell structures. Clearly, as long as the overburden dips in the same direction as the salt floor, the salt must be buoyant for updip extrusion to occur.

To obtain an estimate of possible magnitudes of updip flow rates, consider the mean flow velocity $q/h = (h^2|\Delta\rho|g/12\mu)\partial h_2/\partial x$. For the parameter values $h = 1000\,\mathrm{m}$, $|\Delta\rho|g = 4000\,\mathrm{Pa\,m^{-1}}$, $\mu = 10^{18}\,\mathrm{Pa\,s}$, and an interface slope of 3 degrees ($\partial h_2/\partial x \approx 0.05$), one finds $q/h = 600\,\mathrm{m\,Ma^{-1}}$. This suggests that there may indeed be a significant tendency for large-scale updip migration of salt in areas affected by differential subsidence of the salt floor (Jenyon, 1985).

3.3 Stability and gravitational failure of the overburden; some approximate results

Although it would have been straightforward to retain the first term in expression (10) in the above analysis of kinematic waves, this would have raised the question of the unknown boundary velocities, in particular of $U_2(x,t)$, which must in general be determined simultaneously with the deformation of the surrounding strata. For an understanding of structures in the sedimentary overburden, it is none the less necessary to address this problem of coupled deformation of overburden and substratum. As will be shown in this section, some progress can be made in this direction by rather simple means, up to a point where certain principal modes of overburden response can be understood in quantitative terms.

The problem was first tackled by Last (1988), who used finite-difference methods to solve the problem of simultaneous Maxwellian viscoelastic flow in the substratum and Coulomb-plastic deformation of the overburden. The overburden slope was maintained constant by applying a sediment redistribution condition. In his study, Last not only confirmed the applicability of lubrication theory (Lehner, 1977) for determining the evolution of the overburden/substratum boundary, but also succeeded in rationalizing the numerically predicted overburden behaviour, in terms of simple limit-equilibrium considerations, as a slope failure phenomenon brought about by deep-seated creep.

It will be shown here that limit-equilibrium considerations, essentially borrowed from soil mechanics, can also yield estimates of the overburden displacement rate $U_2(x,t)$. Since overburden extension ($\partial U_2/\partial x > 0$) or compression ($\partial U_2/\partial x < 0$) is associated with corresponding structural styles, some insight may thus be gained into factors that control the appearance in space and time of characteristic structures, such as normal (growth) faults or thrust faults.

Consider again a (subaqueous) wedge of sediments that is building out over a substratum, assuming the interface between overburden and substratum to be horizontal at the current instant in time. It is further assumed that the pore fluid pressure is in hydrostatic equilibrium throughout the overburden. The equations of equilibrium for the overburden, written in terms of the *effective stress* $\sigma'_{ij} = \sigma_{ij} + p\delta_{ij}$, $(i, j, = x, z)$, are then given by

$$\frac{\partial \sigma'_{xx}}{\partial x} + \frac{\partial \sigma_{xz}}{\partial z} = 0, \qquad \frac{\partial \sigma_{xz}}{\partial x} + \frac{\partial \sigma'_{zz}}{\partial z} = \rho'_o g, \tag{45}$$

where $\rho'_o = \rho_o - \rho_w$ as before.

For a homogeneous overburden that satisfies a Mohr-Coulomb yield criterion, there exists a particularly simple pair of solutions to (45) known as *active* and *passive Rankine states* (see, e.g., Mandl, 1988) and the textbook literature on soil mechanics). These solutions have the property that all stress derivatives parallel to the layer surface vanish. For a horizontal layer, free of effective tractions on its surface, the Rankine state solutions to (45) are

$$\sigma'_{xx} = f(z), \qquad \sigma'_{xz} = 0, \qquad \sigma'_{zz} = \rho'_o g(z - h_3), \tag{46}$$

where the function $f(z)$ is determined from the additional requirement that the state of stress must also satisfy a Mohr-Coulomb yield criterion. Since σ'_{xx} and σ'_{zz} are clearly principal stresses, the Mohr-Coulomb criterion for a cohesionless material, expressed in terms of these stresses, reads

$$\sigma'_{zz} - \sigma'_{xx} = \pm(\sigma'_{zz} + \sigma'_{xx}) \sin \phi, \tag{47}$$

where ϕ denotes an angle of internal friction. Since σ'_{zz} is fixed by (46c), there are two Rankine limit states, corresponding to the solutions

$$\sigma'_{xx} = k\rho'_o g(z - h_3), \tag{48}$$

with

$$k = \begin{cases} k_a = (1 - \sin \phi)/(1 + \sin \phi) \text{ in the } \textit{active} \text{ state}, \\ k_p = (1 + \sin \phi)/(1 - \sin \phi) \text{ in the } \textit{passive} \text{ state}, \end{cases} \tag{49}$$

k_a and k_p $(= k_a^{-1})$ being known as the coefficients of active and passive earth pressure, respectively. As is well known, the possible modes of failure associated with these limit states correspond to normal faulting and thrust faulting in Anderson's classification[8] (Mandl, 1988). Of interest here is the fact that (48) together with (49) should yield useful approximate constraints on the minimum and maximum horizontal stress, as long as the boundary conditions demanded for the existence of simple Rankine states are also fulfilled in an approximate sense.

---

[8] It is tacitly assumed here that the intermediate principal stress is oriented perpendicular to the x, z-plane.

To see this, let us first integrate equation (45a) over the overburden thickness, making use of the boundary condition of vanishing effective tractions along the sea bottom:

$$\frac{d}{dx}\int_{h_2}^{h_3(x)} \sigma'_{xx}\, dz = \sigma_{xz}(h_2). \tag{50}$$

The shear stress $\sigma_{xz}(h_2$ can be determined from the boundary condition along the overburden/substratum interface, through which it is equated the shear stress generated by the squeeze flow in the substratum. The latter is found by integrating relation (33), subject to the boundary conditions $v_x(h_1) = 0$ and $v_x(h_2) = U_2$, giving

$$\sigma_{xz}(h_2) = \sigma_o t_0^{\frac{1}{n}} |U_2/h|^{\frac{1-n}{n}} U_2/h + \frac{h}{2}\frac{\partial p}{\partial x}. \tag{51}$$

Substituting $\partial p/\partial x = \rho'_o g \partial h_3/\partial x = \rho'_o g \partial H/\partial x$ for the pressure gradient and entering the resulting expression for $\sigma_{xz}(h_2)$ in (50), one obtains

$$\frac{d}{dx}\int_{h_2}^{h_3(x)} \sigma'_{xx}\, dz = \sigma_o t_0^{\frac{1}{n}} |U_2/h|^{\frac{1-n}{n}} U_2/h + \frac{1}{2}\rho'_o g h \frac{dH}{dx}. \tag{52}$$

Note that the two contributions to the basal shear stress (51), the first deriving from the Couette-flow and the second from the Poiseuille-flow component, will typically have opposite signs, due to the fact that $U_2 > 0$ and $dH/dx < 0$ as a rule. The first component mobilizes the resistance to shear in the substratum, thereby diminishing the build-up of compression in the down-slope direction, while the second one adds to this build-up by 'dragging' the overburden in the direction of the squeeze flow.

Equation (52), in combination with the limit state solutions (49), will now serve to establish three fundamental modes of overburden response.

Stability of a horizontally constrained overburden. In much of the foregoing the overburden layer was assumed to cover the entire substratum and it was therefore treated as horizontally constrained ($U_2 = 0$). This will often be true already in a relatively early stage of development of a salt basin. If the overburden thickness displays a systematic trend, a question of stability then arises with the presence of a mobile substratum leading a build-up in horizontal stress in the down-dip direction. It is here that the notion of Rankine limit stress states offers a simple means for assessing the situation. For if the slope angle, as is usually the case, remains sufficiently small everywhere, it is reasonable to assume that the attainable limit stress states in a Coulomb-plastic overburden will agree closely with the *active* and *passive* states.

One is thus lead to consider the distance necessary for a build-up in horizontal stress from an active (extensional) to a passive (compressional) state. If the total length of the overburden/substratum contact in the down-slope direction exceeds that distance, the overburden is expected to become unstable, a condition that would manifest itself in the development of normal faults on the upper slope and thrust faults at a certain minimum distance down the slope[9]. To answer this question with the aid of equation (52), the velocity U_2 is set equal to zero so as to represent an overburden at rest. The equation is then integrated between the limits x_0 and x, assuming that the state of stress in the overburden at x_0 has attained the *active* limit. Making use of (49) at x_0, one obtains

$$P'(x) \equiv -H^{-1}(x) \int_{h_2}^{h_3(x)} \sigma'_{xx}\, dz = \frac{1}{2}\rho'_o g H^{-1}(x) \left(k_a H_0^2 - \int_{x_0}^{x} h(x') \frac{dH}{dx'}\, dx' \right) \tag{53}$$

for the thickness-averaged effective horizontal pressure. Here $H_0 = H(x_0)$. This pressure must now be compared with the maximum attainable, *passive* pressure in any location x, which is

$$P'_{\mathrm{p}} = \frac{1}{2} k_p \rho'_o g H. \tag{54}$$

Since the ratio P'/P'_{p} cannot exceed 1, by assumption, the condition

$$P'(x)/P'_{\mathrm{p}}(x) = 1, \tag{55}$$

when solved for x, for given functions $h(x)$ and $H(x)$, yields the position $x = x_{\mathrm{c}}$ at which the overburden may be expected to fail in compression by thrust faulting. The distance $x_{\mathrm{p}} - x_0$ also represents the critical length of a finite substratum, beyond which its overburden must be expected to fail as a result of the downslope build-up of horizontal stress.

If the substratum thickness differs locally only little from an average thickness $\bar{h}$, the integral in (53) evaluates simply to $\bar{h}[H(x) - H_0]$. Condition (55) reduces to

$$k_a H_0^2/H^2 + (\bar{h}/H_0)(H_0/H)(H_0/H - 1) = k_a^{-1}$$

in this case, which may be solved for $H_{\mathrm{c}} \equiv H(x_{\mathrm{c}})$, giving

$$H_{\mathrm{c}} = k_a \left\{ \left[\left(\frac{\bar{h}}{2H_0} \right)^2 + \frac{\bar{h}}{k_a H_0} + 1 \right]^{1/2} - \frac{\bar{h}}{2H_0} \right\} H_0. \tag{56}$$

Note that $H_{\mathrm{c}} \rightarrow k_a H_0$ as $\bar{h} \rightarrow 0$. This limit corresponds to a vanishingly small contribution of the Poiseulle flow component to the basal shear stress (51).

---

[9] The first mode of instability could of course be buckle folding rather than thrust faulting. Evidence for this may be seen in the Mexican Ridges fold belt in the western Gulf of Mexico, which has been interpreted as the downdip compensation for updip extension by Worrall & Snelson (1989), although these authors remain in doubt about the presence of salt or other ductile materials in a basal detachement layer. Further evidence for gravitationally induced folding has been documented by Cobbold & Szatmari (1991) and Cobbold *et al.* (1995) for some deep sections of Brazilian continental margin deposits; these data have been compared with theoretical predictions by Triantafyllidis & Leroy (1997).

Given the function $H(x)$ that represents the overburden thickness profile, the position x_c may now be determined. Consider, for example, the function

$$H = (H_0 - H_\infty)e^{-(x-x_0)^2/l^2} + H_\infty, \tag{57}$$

where l is a characteristic decay length. Its inverse yields

$$x_c = x_0 + l\left[\ln\frac{H_0 - H_\infty}{H_c - H_\infty}\right]^{1/2} \tag{58}$$

after substitution of H_c for H from (56). Note, however, that a solution will exist only as long as $H_c > H_{\text{inf}}$. In other words, the overburden will nowhere attain a passive limit state of stress and will thus remain horizontally confined, if $H_c \leq H_{\text{inf}}$.

In general, if the substratum thickness h varies significantly with x, condition (55) will provide an implicit relation for x_c, which is best solved graphically for specific functions $h(x)$ and $H(x)$.

Overburden failure caused by a substratal creep. Let us now examine the situation in which the substratum extends over a distance larger than $x_c - x_0$, so that the overburden is likely to be set in motion, with $U_2 > 0$. In particular, let us consider certain inferences that may be based solely upon equation (52) and an assumption of rigid-plastic behaviour of the overburden. The substratum thickness is again taken to be approximately uniform and equal to h, while the overburden profile at a given instant in time is that of (57), the position x_0 corresponding to the shoreline and the land surface remaining level behind that line. Given that $H_c > H_\infty$, on the basis of the previous static argument, it is now postulated that there exists a current pattern of incipient failure, extending in a down-slope direction from the shore line $x = x_0$, where *active* failure is assumed to occur first.

The occurrence of active failure in an updip position will set the overburden in motion, i.e., U_2 will become positive over some distance in the down-slope direction. In fact, not only U_2, but also its derivative dU_2/dx will be positive, if the overburden is to undergo extension. A simple argument, based on equation (52), would therefore suggest that an active state of stress will prevail in a down-slope direction for as long as the condition $dU_2/dx > 0$ is satisfied and up to the point where $dU_2/dx = 0$ for the first time. To find this point, the active limit value of $\sigma'_{xx} = k_a\rho'_o g(z - h_3)$ is substituted in (52), giving

$$(2\sigma_o t_0^{\frac{1}{n}}/\rho'_o g)|U_2/h|^{\frac{1-n}{n}}U_2/h = -d(k_a H^2 + hH)/dx \tag{59}$$

as an equation for $U_2(x)$ in terms of the known function $H(x)$, which holds throughout the actively failing segment of the overburden. For the thickness distribution (57), the term on the right-hand-side attains its maximum value at a position $x = x_a$. Beyond this point, the assumption of overburden spreading by continuing active failure ceases to be consistent with the overburden gradient;

the compressive horizontal stress is therefore expected to rise above the active limit, leaving the overburden in a 'rigid' state. In other words, having attained a maximum $U_2^{\max}$ the translation velocity will remain constant from at x_a onwards up to $x = x_\mathrm{p}$, where the horizontal effective stress finally reaches the passive limit. Having determined x_a and $U_2^{\max}$ in the first step, the length of the 'rigid block' $x_\mathrm{p} - x_\mathrm{a}$ may be found by integrating (52) between x_a and x_p for $U_2 = U_2^{\max} > 0$, assigning the active respectively passive limit to the horizontal stress in these points. Accordingly, x_p is determined from the implicit relation

$$\frac{k_p H_\mathrm{p}^2 - k_a H_\mathrm{a}^2 - h[H_\mathrm{a} - H_\mathrm{p}]}{x_\mathrm{p} - x_\mathrm{a}} = \frac{2\sigma_0}{\rho_o' g}\left(\frac{t_0 U_2^{\max}}{h}\right)^{\frac{1}{n}}, \tag{60}$$

with $H_\mathrm{a} = H(x_\mathrm{a})$ and $H_\mathrm{p} = H(x_\mathrm{p})$, the function $H(x)$ being given by (57). These steps have been carried out for an example, assuming the parameter values $x_0 = 0$, $H_0 = 5000\,\mathrm{m}$, $H_\infty = 500\,\mathrm{m}$, $h = 500\,\mathrm{m}$, $l = 10^5\,\mathrm{m}$, $\mu = \sigma_0 t_0 = 10^{18}\,\mathrm{Pa\,s}$, and $\rho_o' g = 10^4\,\mathrm{Pa\,m^{-1}}$, $k_a = 0.3$, $n = 1$. The results may be read from Figure 5, which contains plots of the overburden profile $H(x)$ and velocity $U_2(x)$. The ascending branch of U_2 is computed from (59) up to $U_2^{\max} = 7965\,\mathrm{m\,Ma^{-1}}$ at $x_\mathrm{a} = 53906\,\mathrm{m}$ ($H_\mathrm{a} = 3865\,\mathrm{m}$). At that point, the active state is assumed to is assumed to terminate, since its continuation would require U_2 to decline, a condition deemed inconsistent with active extension. U_2 is threfore assummed *not* to decline, but to remain constant and equal to $U_2^{\max}$ up to $x_\mathrm{p} = 138516\,\mathrm{m}$ ($H_\mathrm{p} = 1160\,\mathrm{m}$), which is found from (60). A third, descending branch of U_2 at values of x larger than x_p has been computed from equation (59) for $U_2 = U_2^{\max}$, with k_p replacing k_a based on the assumption of a continuing *passive* limit state. The velocity distribution is consistent with horizontal contraction for this branch. Note, however, that the graph of U_2 fails to intersect the horizontal line $U_2 = U_2^{\max}$ in exactly x_p; the assumption of a smooth overburden profile at x_p would in fact have to be relaxed in order to achieve this match. But there is little to be gained from such an *ad hoc* adjustment in view of the approximate nature of the whole argument. Rather, it is suggested here to compute the values of x_a and x_p as rough indications for the subdivision of the slope into the three structural domains of active extension, rigid sliding, and passive contraction. It should be clear also that these predictions are to be understood in a broad qualitative sense and that the true extent of the three domains will depend strongly on the precise geometry of any real problem as well as on the distribution of material properties etc. Indeed, the main use of an argument as the present one should be in making rough estimates and providing quantitatively plausible explanations for observed or expected phenomena.

Note also that $x_\mathrm{c} = 116482\,\mathrm{m}$, as obtained from (58), is indeed smaller than the value of x_p, the expected difference being due to the basal shear resistance that is mobilized by the gliding overburden.

The overall partitioning of an unstable slope into zones of active failure, rigid gliding, and passive failure agrees with the principal, large-scale pattern that is typically found in prograding delta systems (Evamy *et al.* 1978; Worrall & Snelson, 1989). As has been emphasized, the picture drawn in Figure 5 represents

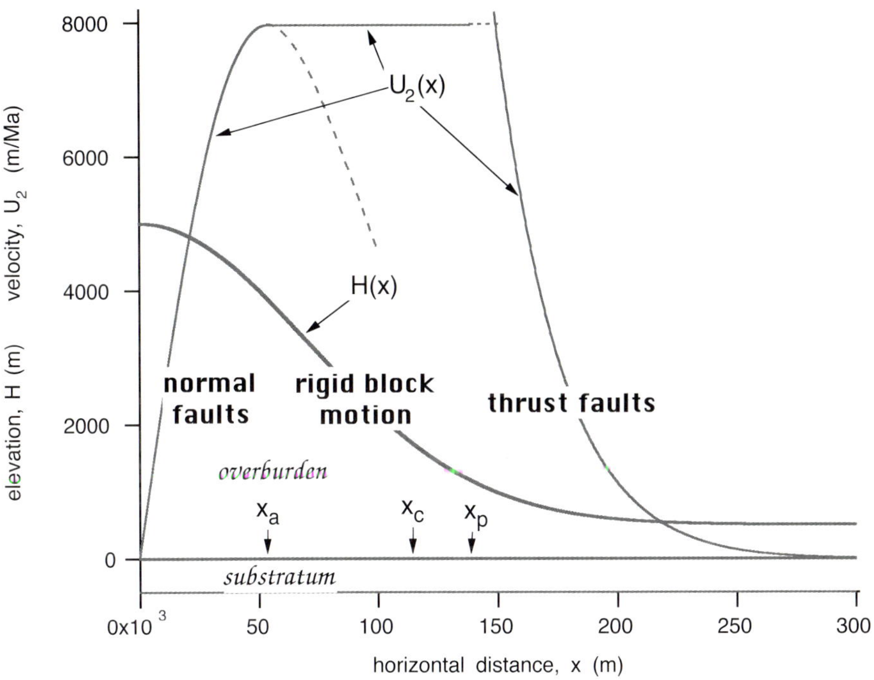

Fig. 5. Gravity-induced failure of the overburden of a mobile substratum

the situation at a given instant in time. The complex structural evolution of any natural setting is in fact the result of a continuous downdip displacement of the failure regimes identified in Figure 5. Since each regime will remain active over a period of time whose length is determined by the rate of progradation of the shoreline, the relative magnitude of the progradation velocity in relation to the average flow rate in the substratum must have a decisive influence on the 'structural style' in the overburden. The progradation velocity is of course determined in part by the local displacements of the sediment at the sea floor and is therefore coupled to the flow in the substratum. Nevertheless, for the purpose of a dimensional analysis, a hypothetical propagation velocity V can be used, which is obtained by shifting an assumed sea bottom profile by an increment Δx_0 per unit of time, the increment being calculated from a total rate of sediment supply as the fundamental independent parameter. The dimensionless quantity relating this progradation speed to the mean flow rate in the substratum is thus given by Vh/q.

Focussing now on the region of active failure, one expects that there will be little time for overburden extension, if $Vh/q \gg 1$; the resulting mode of active failure should thus be characterized by a sequence of densely spaced normal faults with relatively small throws. In the opposite case, there will be enough time for substantial overburden extension to develop and thereby to accomodate a large volume of sediment in depocenters. The latter are overrun by the shore line only when local subsidence rates have slowed down sufficiently by the thinning of the substratum. Such depocenters may typically be bounded by growth faults at their updip and downdip limits, the latter being the site of a major 'counterregional' (i.e., landward dipping) fault, featuring salt (or shale) ridges in the footwall (Verrier & Castello Branco, 1972; Worrall & Snelson 1989; Evamy *et al.*, 1978).

In the above example the instantaneous mean flow rate (10) becomes $q/h = U_2/2 - (\rho_o' g h^2/12\mu)\partial H/\partial x$. At $x = x_\mathrm{a}$ (where $\partial H/\partial x = -2x_\mathrm{a}(H_\mathrm{a} - H_\infty)/l^2 = 0.036$), one finds $U_2/2 = U_2^{\max}/2 = 3980\,\mathrm{m\,Ma^{-1}}$ and $-(\rho_o' g h^2/12\mu)\partial H/\partial x = 227\,\mathrm{m\,Ma^{-1}}$, i.e., the Couette flow component associated with active overburden extension dominates the Poiseuille flow component.

4 Concluding discussion

The theory presented in this paper can contribute to a better qualitative and quantitative understanding of the 'lubrication flow' of salt and mobile overpressured shales or other ductile substrata, driven by differential loads and/or buoyancy forces. In particular, the recognition and quantitative description of a wave-type propagation of substratal *isopachs* by differential loads should be of considerable help in the interpretation of certain geological structures.

The principle limitations of the present theory are those of lubrication theory as well as the assumption of negligible overburden resistance to deformation. For buoyant substrata, the latter leads to a mathematically ill-posed problem of unstable growth of interfacial perturbations, but there appears to be some scope here for simple models that would account for overburden resistance in an approximate manner so as to generate a well-posed problem. In fact, any simple foundation-type model would suffice for this purpose, the only forseeable complication arising from the need to account for sedimentation and/or erosion at the sea floor. A more ambitious continuation would be modelling the overburden as an elastoplastic material as in Last (1988) and Massin *et al.* (1996), but any attempt in this direction would have to come to grips with the problem of faulting, i.e., of localized deformation. Unfortunately, theoretical work in this direction appears to be insufficiently advanced to form a viable alternative to physical analogue models.

From the point of view of a geological 'user', the present theory might also be of some help in suggesting and designing future systematic analogue experiments. For example, the variation in structural style seen in prograding deltaic sediments may well be governed to a large extent by the dimensionless ratio Vh/q of outbuilding velocity over the mean velocity of a substratal squeeze flow. To test such a hypothesis, one might carry out experiments with a wedge of sediments (sand) prograding over a ductile substratum (silicon putty) at different speeds V, such that the dimensionless ratio $V\mu/\rho_o' g l^2$ would be allowed to vary over a significant range - l being a characteristic length of the system, ρ_o' the bulk density of the overburden (dry sand), and μ the viscosity of the substratum (silicon putty). (Note that the ratio $V\mu/\rho_o' g l^2$ is a fully controllable equivalent of the dimensionless parameter Vh/q.) A different experiment might focus on the process of large-scale updip extrusion of buoyant salt for certain types of basin subsidence and infill histories. In each case the insight gained from the present theory can thus serve to motivate a particular experimental design and strategy.

References

Bird, P. (1991). Lateral extrusion of lower crust from under high topography, in the isostatic limit. *J. Geophys. Res.*, **96**, 10,275–286.

Bornhauser, M. (1958). Gulf Coast tectonics. *Am. Ass. Petrol. Geol. Bull.*, **42**, 339–370.

Brown, R.G. (1969). Modern deformational structures in sediments of the Coorong Lagoon, South Australia. *Spec. Publs. geol. Soc. Aust.*, **2**, 237–241.

Buck, W.R. (1991). Modes of continental lithospheric extension. *J. Geophys. Res.*, **96**, 20,161–178.

Carter, N.L., Horseman, S.T., Russel, J.E., and Handin, J. (1993). Rheology of rock salt. *J. Struct. Geol.*, **15**, 1257–1271.

Cobbold, P.R. and Szatmari, P. (1991). Radial gravitational gliding on passive margins. *Tectonophysics*, **188**, 249–289.

Cobbold, P.R., Szatmari, P., Demercian, L.S., Coelho, D., and Rossello, E.A. (1995). Seismic and experimental evidence for thin-skinned shortening by convergent radial gliding on evaporites, deep-water Santos Basin, Brazil. In: Salt tectonics: a global perspective, edited by M.P.A. Jackson, D.G. Roberts and S. Snelson, Am. Ass. Petrol. Geologists Memoir 65, Chap. 14, pp. 305–321.

Evamy, B.D., Haremboure, J., Kamerling, P., Knaap, W.A., Molloy, F.A. and Rowlands, P.H. (1978). Hydrocarbon habitat of Tertiary Niger Delta. *Am. Ass. Petrol. Geol. Bull.*, **62**, 1–39.

Hopper, J.R. and Buck, W.R. (1996). The effect of lower crustal flow on continental extension. *J. Geophys. Res.*, **101**, 20,175–194.

Humphris, C.C. (1978). Salt movement on the continental slope, Northern Gulf of Mexico. In: *Framework, Facies, and Oil-Trapping Characteristics of the Upper Continental Margin*, edited by A.H. Bouma, G.T. Moore and J.M. Coleman, pp. 69–86, Am. Ass. Petrol. Geologists Studies in Geology, No. 7.

Humphris Jr., C.C. (1979). Salt movement on the continental slope, Northern Gulf of Mexico. *Am. Ass. Petrol. Geol. Bull.*, **63**, 782–798.

Jackson, M.P.A. and Talbot, C.J. (1992). Advances in salt tectonics. In: *Continental Deformation*, edited by P. L. Hancock, Chap. 8, pp. 159–179, Pergamon Press, Oxford.

Jackson, M.P.A. and Vendeville, B.C. (1994). Regional extension as a geologic trigger for diapirism. *Geol. Soc. Am. Bull.*, **106**, 57–73.

Jackson, M.P.A., Vendeville, B.C., Schultz-Ela, D.D. (1994). Structural dynamics of salt systems. *Ann. Rev. Earth Planet. Sci.*, **22**, 93–117.

Jenyon, M.K. (1985). Basin-edge diapirism and updip salt flow in the Zechstein of the southern North Sea.*Am. Ass. Petrol. Geol. Bull.*, **69**, 53–64.

Kupfer, D.H. (1974). Environment and intrusion of Gulf Coast salt and its probable relationship with plate tectonics. In: *4th Symposium on Salt*, edited by A.H. Coogan, pp. 197–213, N. Ohio Geol. Soc., Cleveland, Ohio.

Langlois, W.E. (1964). *Slow Viscous Flow*, Chap. IX, Macmillan, New York.

Last, N.C. (1988). Deformation of a sedimentary overburden on a slowly creeping substratum. In: *Numerical Methods in Geomechanics*, edited by G.S. Swoboda, pp. 577–585, Balkema, Rotterdam.

Lehner, F.K. (1973). Structures caused by squeeze flow of a ductile layer under varying overburden. Unpublished company report, Shell Research, Rijswijk.

Lehner, F.K. (1977). A theory of substratal creep under varying overburden with applications to tectonics. *EOS Abstr.*, 508, June 1977.

Leroy, Y.M. and Triantafyllidis, N. (1996). Stability of frictional, cohesive layer on a viscous substratum: variational formulation and asymptotic solution. *J. Geophys. Res.*, **101**/B8, 17795–811.

Lighthill, M.J. and Whitham, G.B. (1955). On kinematic waves. I. Flood movement in long rivers. *Proc. Roy. Soc. London A*, **229**, 281–316.

Mandel, J. (1962). Essais sur les modèles réduits em mécanique des terrains. Etude des conditions de similitude. *Revue de l'Industrie Minérale*, **44**, 611–620. Engl. transl. in: *Int. J. Rock Mech. Min. Sci.*, **1**, 31–42, 1963.

Mandl, G. (1988). *Mechanics of Tectonic Faulting*, Elsevier, Amsterdam.

Mandl, G. and Crans, W. (1981). Gravitational gliding in deltas. In: *Thrust and Nappe Tectonics*, edited by K.R. McClay and N.J. Price, pp. 41–54, Geol. Soc. London Spec. Publ. No. 9, London.

Massin, P., Triantafyllidis, N. and Leroy, Y.M. (1996). Stability of a density-stratified two-layer system. *C.R. Acad. Sci. Paris*, **322**, série II a, 407–413.

Merki, P. (1972). Structural geology of the Cenozoic Niger delta. Proc. Ibadan Univ. African Geol. Conf., Ibadan, Nigeria, pp. 635–646.

Morgan, J.P., Coleman, J.M. and Gagliano, S.M. (1968). Mudlumps: diapiric structures in Mississippi delta sediments. In: *Diapirism and Diapirs*, edited by J. Braunstein, pp. 145–161, Am. Ass. Petrol. Geol. Memoir No. 8.

Nettleton, L.L. (1955). History of concepts of Gulf coast salt-dome formation. *Am. Ass. Petrol. Geol. Bull.*, **39**, 2373–83.

Nye, J.F. (1963). On the theory of the advance and retreat of glaciers. *Geophys. J. r. astr. Soc.*, **7**, 431–456.

Paterson, W.S.B. (1969). *The Physics of Glaciers*, (2nd ed.), Pergamon Press, Oxford.

Rettger, R.E. (1935). Experiments on soft rock deformation. *Am. Ass. Petrol. Geol. Bull.*, **19**, 271–292.

Triantafyllidis, N. and Leroy, Y.M. (1997). Stability of a frictional, cohesive layer on a viscous substratum: Validity of asymptotic solution and influence of material properties. *J. Geophys. Res.*, **102**/B9, 20551–570.

Trusheim, F. (1960). Mechanism of salt migration in Northern Germany. *Am. Ass. Petrol. Geol. Bull.*, **44**, 1519–40.

Vendeville, B.C. and Jackson, M.P.A. (1992). The rise of diapirs during thin-skinned extension. *Mar. Petrol. Geol.*, **9**, 331–353.

Verrier, G. and Castello Branco, F. (1972). La fosse Tertiaire et le gisement de Quenguela-Nord (Bassin du Cuanza). *Rev. I.F.P.*, **27**/1, 51–72.

West, D.B. (1989). Model of salt deformation on deep margin of central Gulf of Mexico Basin. *Am. Ass. Petrol. Geol. Bull.*, **74**, 1472–82.

Whitham, G.B. (1974). *Linear and Nonlinear Waves.* Wiley-Interscience, New York.

Worrall, D.M. and Snelson, S. (1989). Evolution of the northern Gulf of Mexico, with emphasis on Cenozoic growth faulting and the role of salt. In: *The Geology of North America - an Overview*, edited by A.W. Bally and A.R. Palmer, pp. 97–138, Geol. Soc. Am., Boulder.

Fractal Concepts and their Application to Earthquakes in Austria

Wolfgang A. Lenhardt

Central Institute for Meteorology and Geodynamics/Department of Geophysics, Vienna, Austria

Abstract. Fractal concepts have been applied widely in different scientific disciplines since their introduction. This paper discusses practical aspects of determining fractal dimensions, and their interpretation in terms of seismology. Firstly, several evaluation algorithms are investigated. Austria's borderline is then used as an example for demonstrating the effect scale length can have on perimeter estimates. Derived fractal dimensions D concentrate at 1.18, but can vary by more than one decimal of a unit, depending on the algorithm employed. Comparing fractal dimensions across the literature and their interpretation in terms of underlying mechanisms is therefore extremely difficult.

Applying the fractal concept to earthquake statistics reveals that several factors need to be considered, thus leaving room for a variety of interpretations. The common approach estimates D of earthquakes from the b-value - the ratio of the number of small to large earthquakes. The relation between D and the b-value is not necessarily straightforward, but may deviate due to variations in stress drops of earthquakes, stress environments, the geometrical nature of rupture planes and the fault interactions. Therefore, further interpretations concentrated rather on the b-value than on the fractal dimension D.

Most regions in Austria exhibit b-values near unity, which implies normal stress drops and three-dimensional deformation processes. It is shown, that areas of low b-values do not necessarily call for alternative deformation processes, but rather for lower horizontal stresses or higher stress drops.

1 Introduction

The concept of fractals has been successfully applied in explaining natural processes that exhibit chaotic behaviour. In this context, chaos is defined as system evolving from deterministic equations, whose adjacent solutions tend to diverge exponentially in phase space (Turcotte, 1992). In other words, the process can be understood as a non-predictable response of a very complex - but completely defined - system. Thus, very small changes on the input-side of that system can alter the output in a virtually erratic way. Even after considering all known and possibly involved processes, we end up with a system response which behaves chaotically. Such a system has been referred to as *deterministic chaos* by Huang and Turcotte (1990).

Earthquakes can be understood as an expression of such a chaotic system which involves many factors. Hence, earthquake prediction is mainly unsuccessful because too many parameters - such as tectonic stresses, fault geometry's, local differences in cohesion, etc. - are essential input parameters, which are difficult to ascertain in order to permit a useful earthquake prediction.

However, this paper does not deal with issues of prediction, but tries to shed some light on earthquake occurrences on a broader scale. For instance, the factor of magnification or *scale*, which enables one to investigate a problem in greater detail, can have an influence on the final result. These so-called *scaling laws* can also bear information on earthquake mechanisms and earthquake hazard (Turcotte, 1989).
These laws are of paramount importance for

- distinguishing different processes at a particular scale,
- predicting the behaviour of a system across a wide range of input-data.

The constants involved in these scaling-laws can be related to *fractal dimensions* of tectonic systems - and have increasingly attracted the interest of scientists since the 70's. The first part of this paper deals with different fractal approaches and their applications. Later on, fractal concepts will be applied to earthquake data from Austria.

2 Fractal concepts

The idea of fractal - non-integer - dimensions became widely famous through Mandelbrot (1982). Based on Richardson (1961), Mandelbrot demonstrated, that the relation between different scales r and perimeters P of coastlines follows:

$$P = C_1 * r^{-C_2} \tag{1}$$

with C_1 and C_2 being constants. Graphs, displaying this data in a log-log fashion are named *Richardson-plots*, because both the basic idea and plotting the data in this way was started by Richardson. Such graphs permit the determination of the *fractal dimension* D, because the perimeter and the fractal dimension are linked to each other. If a certain number N_i of elements of length r_i is needed to describe an object,

$$N_i = \frac{C}{r_i^D} \tag{2}$$

with C being a constant, than the estimate of the perimeter P_i is given by

$$P_i = r_i * N_i \tag{3}$$

Substituting (eq.2) into (eq.3) gives

$$P_i = \frac{C}{r_i^{D-1}} \tag{4}$$

or

$$P_i = C * r_i^{1-D} \tag{5}$$

Hence, the slope of this particular relation equals $1 - D$, with D the fractal dimension. The concept of log-log distributions and their meaning in nature can be traced further back to Harold Edwin Hurst, an engineer, who was mainly involved in resolving the problem of determining the optimum size of a water reservoir along the Nile in Egypt and predicting lowest and highest capacities according to annual floods (Hurst, 1951). He managed to formulate the problem in the *re-scaled range analysis*, by relating the total range $R(t)$ of extremes (difference between maximum and minimum value of a time-series of length t) to its standard deviation $S(t)$ by

$$\frac{R(t)}{S(t)} = (a * t)^H \tag{6}$$

Thus, Hurst's formula resembles again a scaling law that incorporates the constants a and H, the latter is being referred to as the *Hurst index*. These constants were found to be $a \sim 0.5$ and $H = 0.72 \pm 0.1$ for most natural processes (Lomnitz, 1994).

Kolmogorov (1941), who described fragmentation processes of rocks, also found a log-log dependence between fragment-size and the number of fragments. Since then, besides other natural phenomena, such as joint roughness and fracture profiles (e.g. Odling, 1994, Seidel and Haberfield, 1995), fragmentation processes have been frequently studied in terms of their fractal behaviour (Turcotte, 1986, Kaye, 1993).

Although this paper deals mainly with self-similar problems, the two terms *self-similar* and *self-affine* should be briefly clarified, before discussing different kinds of evaluation routines:

Self-similar objects are isotropic. The fractality of an object does not depend on the orientation of the axes. The general definition is

$$f(r * x, r * y) \sim f(x, y) \tag{7}$$

with r as the scaling factor.

Self-affine fractals are not isotropic and different co-ordinates are scaled by different factors. As an example, all kinds of time-series such as air-temperatures or strains belong to self-affine systems. The formal definition is (Turcotte, 1992)

$$f(r * x, r^\kappa * y) \sim f(x, y) \tag{8}$$

with κ describing the deviation from self-similarity. Another issue, which needs to be considered, is the difference between deterministic and statistical fractals. Exact self-similar objects, such as the Menger sponge (Fig.1), are referred to as deterministic - or sometimes geometrical - objects. Objects in nature are, however, random - and only self-similar in a statistical way (Barabasi and Stanley,

Fig. 1. The 'Menger' sponge (from Mandelbrot, 1982)

1995). Examples of statistically self-similar objects are coastlines, fragments of rock, etc.

In the following, we quickly look at one important aspect of fractals. Obviously, the fractal dimension of self-similar objects derived from two- and three-dimensional Euclidean space (E) cannot be deduced from each other. A fractal dimension D_E, which has been established from 2D-data ($E = 2$) is not exactly one unit smaller than the fractal dimension of the corresponding 3D-case ($E = 3$) case. Using the same information, the κ-factor (eq.8) cannot be determined, too. As an example, we may consider the Menger sponge (Fig.1) again, which represents a Cantor dust (1D-case, $E = 1$) in three dimensions ($E = 3$). The two-dimensional equivalent ($E = 2$) is called Sierpinski carpet. When calcu-

Table 1. Fractal dimensions D_E of self-similar deterministic objects

E-Euclidean dimension	deterministic object	D_E	$D_E - D_{E-1}$
1	Cantor dust	0.631	-
2	Sierpinski carpet	1.893	1.262
3	Menger sponge	2.727	0.834

lating the individual fractal dimensions of the three related deterministic objects (Tab.1), we find, that the differences between their fractal dimensions are not integers. Hence, even for this idealistic set of deterministic objects, we cannot deduce the corresponding 3D-fractal dimension from a lower order fractal dimension, although this has become custom in literature by adding one or two units (e.g. Hirata, 1989, Turcotte, 1992, Zhang et al., 1995).

2.1 Methods

Recalling (eq.2), the general basic equation of a fractal distribution is

$$N_i = \frac{C}{r_i^D} \tag{9}$$

or

$$\log_{10} N_i = C - D * \log_{10} r_i \tag{10}$$

with N_i = number of objects with a characteristic linear dimension r_i, C is a constant of proportionality and D_E is the fractal dimension; D for short.

The following methods are most frequently used for determining *spatial* fractal dimensions:

Yard stick or Ruler or Equipaced Method This method is used to 'measure' the deviation of an object from a perfect line of fractal dimension $D = 1$. It constitutes the classic approach - made widely famous by the paper *How long is the coast of Great Britain?* by Mandelbrot (1967). Considering Richardson's work (1961), Mandelbrot demonstrated that the stated question cannot be answered, but similarities between different scale lengths and perimeters can be established. A similar approach is called *penny-plating* (Kaye, 1993). The name relates to the counting of touching circles covering a 'linear' object.

Box-counting or Amalgamation and Cube-counting These methods are frequently used, especially for dimensional problems of higher orders, such as fragmentation processes. The calculation is normally aided by computers. With the box-counting method, the number of squares needed, to cover a certain outline or an area, are counted. The same approach can be substantially simplified when estimating areas (Mandelbrot, 1975), by representing each 'box' by a dot in the centre ('dot-counting method').

2.2 How Long is the Border of Austria?

Before applying fractal concepts to seismological issues and interpreting the results, we rephrase Mandelbrot's famous question *How Long is the Coast of Great Britain?* and examine the practical aspects of determining fractal dimensions by applying the stated methods to the borderline of Austria. We know already from

Mandelbrot, that the cited question cannot be answered, although the official length of Austria's border is 2706 km (ÖSZ, 1993).

Firstly, we apply the yard-stick method. Following Mandelbrot (1982), we can calculate the fractal dimension from the perimeter distribution. Defining the estimated perimeter according to (eq.3) and considering (eq.2), we get

$$\log_{10} P_i = C - (D - 1) * \log_{10} r_i \tag{11}$$

for the perimeter estimate Pi (see also eq.5). By stepping along the outline of the chosen object - in this case the border of Austria - we find, that we are unable to reach exactly the starting point again. In this case, the starting point was chosen *arbitrarily* as the most eastern border point of Austria (Fig.2). Hence, we

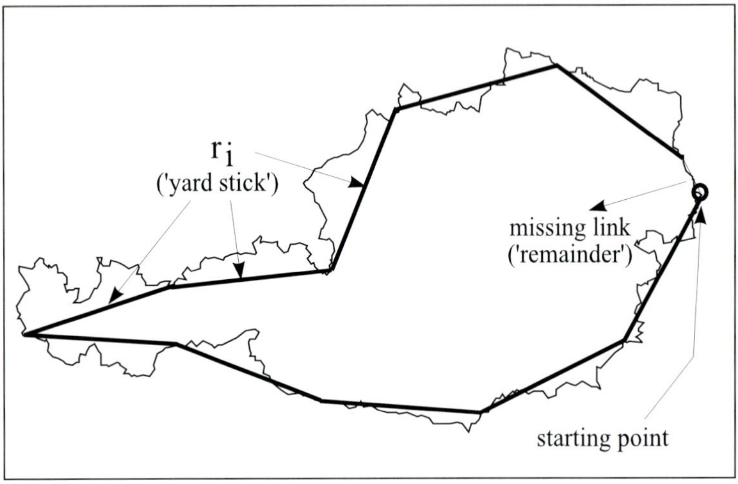

Fig. 2. Principle of the 'yard-stick' method.

can exclude the last full step (technique Y_1), include the last full step (technique Y_2) or we can account for the missing link ('remainder', technique Y_3). With the last method we can improve the perimeter estimate by adding the remainder ε_i, thus

$$P_i = (N_i * r_i) + \varepsilon_i \tag{12}$$

Admittingly, the remainder becomes less important at very small scales. Beyond a critical scale, which is normally defined by the lower limit of the data-resolution, the remainder certainly can have a very disturbing influence on the calculation of fractal dimensions. However, the common approach Y_2 includes the full length of the last 'yard stick'. Hence, the perimeter estimate is exactly one 'yard stick' longer than from the Y_1-technique.

Depending on the employed technique, we end up with different fractal dimensions. The perimeter estimate P (method Y_1) with $N * r$ gives $D = 1.23$ and the conservative perimeter estimate $(N+1)*r$ (method Y_2) yields $D = 1.10$. The third technique Y_3 includes the remainder ($P = (N*r)+\epsilon$) and we get $D = 1.19$ (Tab.2). We note, that the result depends strongly on the technique employed and on how the fractal dimension is calculated in practice. The most stable yard-stick technique seems to be Y_3, which considers the missing link. The classic approach (Y_2) leads to a 10 % smaller fractal dimension. The box-counting

Table 2. Perimeter estimates and fractal dimensions of Austria's border as a function of several 'yard stick' techniques.

				Y_1	Y_2	Y_3
yard stick r (km)	number of yard sticks N	number of yard sticks + 1 $= N+1$	remainder ε in km	$N*r$ in km	$(N+1)*r$ in km	$(N*r)+\varepsilon$ in km
25	65	66	21	1625	1650	1646
50	29	30	11	1450	1500	1461
100	12	13	74	1200	1300	1274
200	6	7	50	1200	1400	1250
400	2	3	133	800	1200	933
fractal dimension (see also Tab.3) $\Rightarrow$				1.23	1.10	1.19

method permits us again, to study the ruggedness of the borderline in terms of the fractal dimension. This time, the data set is given by the box-side length r and the number of boxes to cover the borderline ($r = 25$ km, $N = 86$; $r = 50$ km, $N = 37$; $r = 100$ km, $N = 16$; $r = 200$ km, $N = 5$; $r = 400$ km, $N = 2$). The fractal dimension amounts to 1.29 across the range of scales - and is therefore approximately 10% higher when compared with the yard-stick method. But now we encounter the problem of 'fractal rabbits' (Kaye, 1993), which are strong variations of results from Richardson plots. They occur, when the applied scale approaches the characteristic length of an object. According to Kaye (1993), the largest applied scale should not exceed 1/3 of the characteristic length. Defining the characteristic length as the average of the longitudinal and latitudinal extent of Austria, we find roughly (600km+300km)/2 = 450km. Hence, we should not have used scales larger than 150 km. Hence, results from the box-counting method should be discarded if scales of 200 and 400 km's are included ($D = 1.37$ and $D = 1.35$, see Tab.3). To a certain extent, this phenomenon is also observed in the yard-stick method. Comparing the three techniques Y_1-Y_3 of the yard-stick method, we find that the standard deviations of D become small for scales below 200 km (Tab.3). Accepting $D = 1.18$ as a representative value for the fractal dimension of Austria's border, and considering the regression coefficients from the Richardson plots in the range from 25 to 100 km (intercept = 3.48, slope = −0.18), we can estimate the scale of resolution down to which the official stated length of 2706 km of Austria's border is valid: 1 - 2 km.

Table 3. Fractal dimensions D of Austria's border derived with different methods and techniques. Y refers to the yard-stick method, indices refer to the employed technique: 1= ignoring last incomplete member, 2= including last member, 3= including remainder.

method	$\Leftarrow D \Rightarrow$ (considered range in km's for linear regression)					
$\Downarrow$	(25-400)	(25-200)	(25-100)	(25-50)	mean	std. dev.
Y_1	1.23	1.16	1.22	1.16	1.19	0.04
Y_2	1.10	1.09	1.17	1.14	1.13	0.04
Y_3	1.19	1.14	1.18	1.17	1.17	0.02
box-counting	1.37	1.35	1.21	1.22	1.29	0.08
mean	1.22	1.19	1.19	1.17		
std. dev.	0.11	0.11	0.03	0.03		

2.3 Discussion

Comparing and interpreting fractal dimensions should be carried out with great care. It does not seem advisable to compare fractal dimensions across the literature without knowing exactly, how the fractal dimensions were determined. Comparisons should be restricted to results from identical methods and techniques, for different approaches can lead to different results!

This can be further illustrated by comparing results quoted by different authors. Mandelbrot (1967) states $D = 1.25$ for the west coast of Great Britain, whereas Kaye (1993) calculated $D = 1.28$ with a *different technique* (Kaye was mirroring Great Britain's west coast along its central NS-axis, and only then applying the yard-stick method). Kaye's result is interesting, because it indicates, that rugged parts of an object can dominate the fractal dimension of the whole object.

Considering the very same geometrical object, we find in the literature that *different methods* lead to different results, too (see also e.g. Sornette et al., 1993). Kaye used the 'penny-plating' procedure to calculate the fractal dimension of the whole coast of Great Britain, arriving at $D = 1.29$. Applying the yard-stick technique led him to D =1.28, whereas Richardson found $D = 1.24$ (Kaye, 1993).

In addition, care should be taken when interpreting results. Turcotte (1992) states, that topography subject to erosional processes typically exhibit fractal dimensions of $D = 1.25 \pm 0.05$. When looking at Austria's border and results from the yard-stick method, and especially from its 'Y_1- technique in the range of 25 - 400 km, (D = 1.23) and the box-counting method in the range of 25 - 50 km ($D = 1.22$), one can argue, that the shape of Austria's political borderline is being dictated mainly by erosional processes, which are represented by rivers and mountain chains. Considering other techniques, this interpretation is less pronounced.

Reasons for these discrepancies among fractal calculations are manifold. Not only the application of different calculation methods (e.g. 'yard-stick' or 'box-counting') and techniques (e.g. Y_1 or Y_3), but also subjective definitions, such as the starting point, the choice of subdivisions, the largest included scale, the chosen direction of counting (clockwise or anti-clockwise), and methods of calculating the slope affect the final result. Similar problems were encountered, e.g., by Seidel and Haberfield (1995), when they compared fractal dimensions of joint roughness coefficients quoted by different authors.

3 Earthquakes in Austria - an application

This paragraph begins with a brief overview of one fractal application in seismology - the relation between the so-called b-value and its corresponding fractal dimension. Later on, earthquakes in Austria are evaluated and the results are discussed.

3.1 Fractals in Seismology

The relation of the frequency of earthquakes to their magnitude (a logarithmic expression of the seismic energy released by earthquakes)

$$\log_{10} N = a - b * M \tag{13}$$

was first proposed by Gutenberg and Richter (1954), originally defined for surface wave magnitudes M_s. Since then this formula has been widely used in earthquake hazard analysis and also applied to other kinds of magnitude-scales. The constant a constitutes a measure of seismicity. With few exemptions (i.e. Arab et al., 1994), little attention is paid to a for it depends only on the amount of involved data. However, the slope b - commonly referred to as the *b-value* - describes the relation between smaller and larger earthquakes. Lower b-values are sometimes understood as an indicator of a higher earthquake hazard.

A more generalised form of this relationship divides the number of events N by the time-span T - and possibly by the area, where the earthquakes originated. This enables the reader to compare results from different regions and different time-spans. The most common version of the above stated formula is the cumulative frequency-magnitude relation

$$\log_{10} \left[\frac{N(> M)}{T} \right] = a - b * M \tag{14}$$

which describes the number of events *exceeding* a certain magnitude per year. It can be shown, that the determination of the b-value is theoretically independent of the kind of applied statistic. Cumulative or non-cumulative statistics are comparable, as long as we are interested in the slope - the b-value - only (see also Turcotte, 1992). It has been widely accepted, that the frequency-magnitude relationship constitutes a fractal law (Aki, 1981).

Determining the *b*-value When dealing with large sets of data, the *b*-value is commonly estimated with the maximum likelihood method (see also Bullen and Bolt, 1987) by

$$b = \frac{\log_{10}(e)}{(M_{\text{average}} - M_{\text{min}})} \tag{15}$$

The minimum magnitude M_{min} is the smallest - or threshold - magnitude considered, and M_{average} is the average magnitude. The evaluation of the b-value is not as straightforward as it might appear from (eq.15), because M_{min} - this is where the graph starts deviating from a linear trend - must be defined first. Further, M_{min} is not a constant, but will vary spatially and temporarily. Hence, subsets of data (or regional seismic data) need to be evaluated individually in terms of their regional threshold magnitude.

It has been argued, that the maximum-likelihood estimate is superior to a least-squares fit in so far as the latter can be biased by the presence of a few, large, seismic events (see also Bullen and Bolt, 1987). Since no earthquakes of extreme magnitudes formed part of the regional data sets of Austria, and some regions contained less than 50 earthquakes, which are deemed the lower limit when applying the maximum-likelihood method (Utsu, 1965), the least-square method was used. To detect automatically the threshold-magnitude M_{min}, the cumulative frequency distribution was split into two parts. The left part of the graph (below magnitude M_i) should have a zero-slope, a mean and a variance σ_1. The right side of the graph should follow the $\log N = a - b * M$-relation, which is calculated by omitting single magnitude-occurrences on the most right hand side of the distribution, since they should not be used in recurrence statistics. The variance of the right side of the graph is then expressed by σ_2. When covering the range from, for example magnitude 0 to 5, one would start with M_i set to '0' and then let it steadily increase by for example 0.1 while always calculating the sum of $\sigma_1 + \sigma_2 = \sigma_i$. At a certain M_i, the sum σ_i' reaches a local minimum. The corresponding M_i represents the lower magnitude level M_{min} after which the distribution follows the desired linear relationship. Applying this method to earthquakes of Austria, we find (see also Fig.3):

$$\log_{10}[N(> M)/\text{year}] = 3.16 - 0.86 * M \tag{16}$$

The minimum magnitude M_{min}, below which earthquake data from whole Austria start to become increasingly incomplete, is 2.5. Hence, the $\log N - M$ distribution is approximated best by a linear regression above this magnitude. The desired parameter *b*-value or the negative slope of the frequency-magnitude distribution is 0.86 in this case. Later on, we have a look at several seismic regions in Austria and their regional b-values. In order to do this, each regional magnitude-threshold was determined with the technique presented.

Another factor, which can cause a distortion of the $\log N - M$ statistic - and hence the *b*-value - can be the kind of magnitude, on which the statistic was based. Because body-wave magnitudes m_b scale with surface wave magnitudes

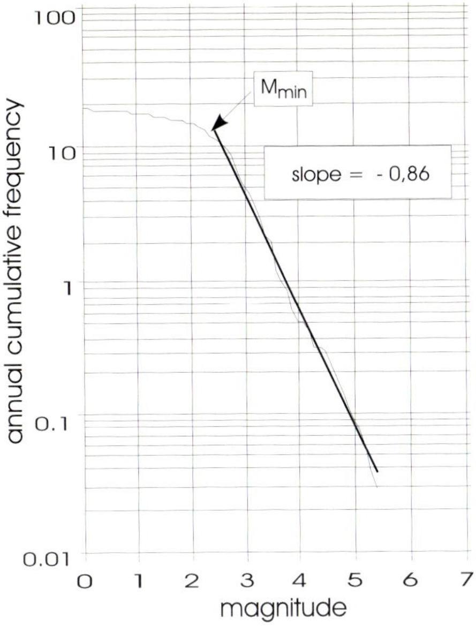

Fig. 3. Cumulative frequency-distribution of magnitudes of Austria.

M_s according to (Richter, 1958)

$$m_b = 2.5 + 0.63 * M_s \tag{17}$$

it becomes obvious, that a b-value based on body-wave magnitudes would tend to exceed b-values deduced from surface-wave magnitude data, especially when large earthquakes are investigated. However, this aspect should not have had an influence on the following b-value-results from Austria, because the largest magnitude earthquakes in Austria can be considered moderate when compared with those of other active regions of the world.

The frequency-magnitude relationship is not the only one that can be expressed in a fractal-like fashion. Intensities of earthquakes follow a similar distribution, although the slope of its frequency graph should not be confused with the b-value derived from magnitudes. Macroseismic intensities (a subjective scale describing human perception of ground shaking as well as damage to buildings, Grünthal, 1993) are an effect of ground motions, which depend on the focal depth and the magnitude of earthquakes. Hence, intensities need to be 're-scaled' for each earthquake, to determine the magnitudes for further evaluation of the *correct* b-value, which could finally be related to the fractal dimension of earthquake processes of a region. The same applies to statistics, which are based on the logarithm of seismic energies. However, in the following discussion, we deal with magnitudes only.

Interpreting the *b*-value Aki (1981) demonstrated, that the statistical b-value of seismic activity can be explained by scaling laws applicable to fault processes. He found that the *fractal dimension amounts to twice the b-value*. Hence, the b-value can be used to calculate the fractality of seismotectonic fault systems. This application is deemed by the author to be extremely important, for some faults do not outcrop ('hidden faults', for example the Northridge earthquake on January 17, 1994), but their fractal characteristics could still be estimated from seismological observations.

Before calculating the b-values of different regions in Austria, and deriving their fractal dimension, a quick look into the involved assumptions seems to be appropriate. Firstly, the magnitude is related to the seismic moment M_o according to

$$\log_{10} M_o = d + c * M \tag{18}$$

with $d = 9.1$ (if M_o is defined in Nm) and $c = 1.5$. The constants were established by Kanamori and Anderson (1975) and Hanks and Kanamori (1979). The moment M_o is defined by

$$M_o = G * A * s \tag{19}$$

with G being the shear modulus and s represents the average slip or displacement along the rupture plane of size A. The shear modulus constitutes a variable on its own, but does not vary a lot in the upper crust, especially in logarithmic terms.

Further, Kanamori and Anderson (1975) found, that the size of the rupture plane A correlates with the moment M_o according to

$$M_o \sim \alpha * A^k \tag{20}$$

with α being a constant and k was found empirically to be $\sim \frac{3}{2}$. This empirical value matches exactly the theoretical concept of Brune (1970, 1971), leading to

$$M_o = \left(\frac{16 * \Delta\tau}{7 * \pi^{\frac{3}{2}}} \right) * A^{\frac{3}{2}} \tag{21}$$

Hence, the constant a includes the stress drop $\Delta\tau$ (see also Lay and Wallace, 1995). The exponent k is therefore well substantiated, and should not vary too much for 'small earthquakes' - that is for magnitudes below 6.6. k-values smaller than 1.5 indicate, that the fault width has become independent of the thickness of the seismogenic layer, where earthquakes originate. The latter applies normally only to 'larger' earthquakes, which rupture across the seismogenic layer - causing the effect, that large earthquakes scale with the square of the fault-length - or area A - rather than with the third power of the fault length - or $A^{\frac{3}{2}}$ (Shimazaki, 1986).

Following Aki (1981), King (1983) and Turcotte (1992), the frequency-magnitude relationship (eq.14) can now be rewritten by substituting the magnitude by its moment (eq.18), and the moment by its rupture area A (eq.20):

$$\log_{10} N = (-k * b * \log_{10} A) + \log_{10} C \tag{22}$$

The constant C is of minor importance, because we are actually interested in the slope in (eq.22). The constant is stated here just for reasons of completeness:

$$C = (b * \frac{d}{c}) + \log_{10} a - \left(\frac{b}{c}\right) * \log_{10} \alpha \qquad (23)$$

Hence, the number of earthquakes scales with the rupture area A thus resembling the basic fractal law

$$N = C * A^{-k*\frac{b}{c}} \Leftrightarrow N = C * r^{-D} \qquad (24)$$

Setting $A \sim r^2$, leads to

$$D = \frac{2 * k * b}{c} \qquad (25)$$

and substituting '1.5' for k

$$D = 2 * b \qquad (26)$$

provided $c = 1.5$. This is essentially the result, Aki (1981) arrived at. Accordingly, the fractal dimension D of earthquakes in Austria would amount to 1.72, based on the b-value of 0.86 (see Fig.3). This slope - and therefore the fractal dimension - can be expected to be different when evaluating individual seismic regions of Austria.

Another important parameter, which has not been discussed yet, is the shear stress drop. The relation between the released seismic energy E_s (in Joule) and the magnitude M (Gutenberg and Richter, 1956, Spottiswoode and McGarr, 1975)

$$\log_{10} E_s = 4.8 + 1.5 * M \qquad (27)$$

the relation between the seismic energy and a complete release of the excess shear stress (Ryder, 1988) - which is commonly referred to as the stress drop $\Delta\tau$

$$E_s = \frac{\Delta\tau * M_o}{(2 * G)} \qquad (28)$$

(see also Kanamori, 1977), and substituting Hanks and Kanamori's relation (eq.18) into (eq.27) and (eq.28), we find that the stress drop amounts to

$$\Delta\tau = G * 10^{-4} \qquad (29)$$

This fact is not surprising, for Kanamori (1977) had assumed the constant stress drop (eq.29), when he and Hanks defined the magnitude-moment relation (eq.18) two years later. This simple stress-drop estimate (eq.29) results in a reasonable value of 3 MPa, given a shear modulus of 30.000 MPa, which is typical for the earth's crust. Hanks and Kanamori's assumption is substantiated by numerous observations, showing that stress drops of most earthquakes vary between 1 and 10 MPa only (see also Scholz, 1982).

A contradiction? Whether or not the spatial earthquake distribution in a defined region is fractal, can either be a result of a fractal distribution of earthquakes along a fault system, or faults of a fault system are statistically distributed in a fractal manner themselves. Because most fault systems exhibit fractal character (King, 1983), the latter could be the case. This point was discussed by Hirata (1989) in great detail when he investigated the actual distribution of epicentres (projection of the earthquakes focus on the earth surface) in space- and time domain - and finding empirically

$$D_H = 2.3 - 0.73 * b \tag{30}$$

At this point, it should be noted, that Hirata (1989) also quoted another relation between D_H and the b-value ($D_H = 1.8 - 0.56 * b$), which indicates strong variations between different data-sets and a generally poor correlation coefficient.

Hirata points out, that the spatial epicentre distribution results in another type of fractal dimension D_H, which must not be confused with Aki's fractal dimension D. Aki (1981) assumed a barrier model, in which smaller earthquakes are a result of a larger earthquake, and deduced the fractal dimension *indirectly* from earthquake-magnitudes. In this case, the fractal dimension is allowed to vary between 2 (defining a plane) and 3 (associated faulting, thus filling a volume). Hirata (1989) refers to Aki's fractal dimension as a *capacity dimension* of a distinct fault process in which earthquakes are related to each other, - whereas the dimension D_H should be understood as a *correlation dimension* (Hirata, 1989) of epicentres associated with a number of faults. According to Hirata, low b-values are then associated with large D_H and large b-values with small D_H. It might seem, that Hirata's empirical findings contradict Aki's theory. This is not necessarily the case, however, when considering the different approaches involved:

Aki *assumed* a single plane along which earthquakes occur, while Hirata calculated the distribution of the distance between *successive* epicentres ('spatial sequence') associated with several faults. Besides, it was demonstrated in the beginning, that adding one unit to Aki's theoretical result (Hirata, 1989), to make up for the deficiency between the epicentre distribution (2D) and Aki's model (3D), cannot be supported from a fractal point of view. In addition, we recognise, that focal depths are not distributed in a similar way as epicentres are. This can be seen clearly in an intra-plate scenario, where most earthquakes occur in the top layers of the earth's crust (Fig.6, or see Scholz, 1982). Therefore, we encounter certainly a *non self-similar* problem when comparing the two approaches of Aki and Hirata. Hence, horizontal and vertical scales cannot be compared and fractal calculations derived from them, should lead to different results.

Another example of such a controversy can be found in the paper of Odling (1994) - who finds a negative correlation between the fractal dimension and the joint roughness coefficient of fracture profiles - and in the paper of Seidel and Haberfield (1995), quoting - among other authors - a positive correlation.

Factors affecting the ratio D/b In Figure 4, the ratio of D/b is plotted against the source geometry factor k as a function of Hanks and Kanamori's c-factor:

$$\frac{D}{b} = 2 * \frac{k}{c} \tag{31}$$

(see also eq.25). Accordingly, we find that

- D is smaller than $2 * b$ - if k is small or c is large, or $\frac{k}{c} < 1$ - and
- D is larger than $2 * b$ - if k is large or c is small, or $\frac{k}{c} > 1$.

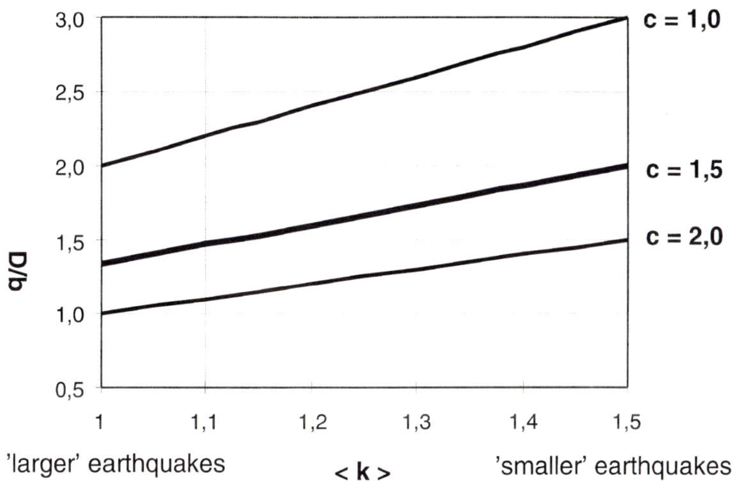

Fig. 4. Relation of D/b as function of the fault geometry factor k and of c.

It was shown previously, that k-values ~ 1.5 apply to intra-plate scenarios, such as the seismotectonic regime prevailing in Austria, whereas k-values < 1.5 are rather applicable to larger earthquakes (M>6.6, see Shimazaki, 1986).

Hanks and Kanamori's c-value (also 1.5) assumes a constant stress drop, however. According to equation (eq.18) and (eq.28), c-values below 1.5 would therefore serve as an indication of higher stress drops - and c-values above 1.5 suggest smaller stress drops, while the fractal dimension of the fault system can remain the same, provided Gutenberg and Richter's relation between released seismic energy and magnitude holds for the investigated range of magnitudes (see eq.27).

Two-dimensional fault processes (fractal dimension D near 1.5) are unlikely to be experienced in nature (King, 1983). But we can still explain low b-values with a three-dimensional deformation, by interpreting the b-values in terms of c or k. To make up for the deficiency between D and the b-value, a smaller

c-value (<1.5) or a larger k-value (>1.5), would be sufficient to allow for three-dimensional faulting processes in areas of low b-values. A small c-value implies higher stress drops during the rupture process in these regions. It represents another likely reason for low b-values, for higher k-values are unlikely to be encountered. Stress drops exceeding standard values (eq.29) by a factor of 3 - which still complies with the range of stress drops observed world-wide - are already sufficient to alter the magnitude scale in such a way, that lower b-values are observed.

This rather lengthy excursion into seismology and fractals demonstrates, that interpreting the fractal dimension of earthquakes is not an easy task. In addition, the fractal dimension D of an earthquake activity amounts to twice the b-value - strictly speaking - *only if*

- magnitudes are used for calculating the b-value,
- the seismic energy is related to the magnitude according to Gutenberg and Richter (1956),
- the seismic moment scales with the rupture area according to Kanamori and Anderson (1975),
- the magnitude is related to the moment according to Hanks and Kanamori (1979)
- and the barrier-model applies (Aki, 1981).

3.2 Regional b-values in Austria

Austria's seismicity can be considered as moderate and may compared with that of Switzerland. Sixteen earthquakes are felt in Austria on average each year, but many more are detected by seismic instruments. Since 1900 forty-five earthquakes have had an epicentral intensity of grade 6, eight earthquakes had an intensity of grade 7 and one event resulted in an intensity of grade 8. The latter tremor caused extensive damage to buildings of a village, just few km's south of Vienna. In Figure 5 all felt earthquakes are plotted, giving an impression of the spatial distribution of seismicity in Austria during the time period from 1900 until 1994. The focal depths of earthquakes in Austria show a dominant peak at 7 km, but vary considerable between 2 and 13 km (Fig.6).

Seismic regions The b-values of fifteen individual seismic regions in Austria will now be discussed. The borderlines of the regions were defined according to seismic and geological information (Lenhardt, 1995). Regional b-values are shown in Figure 7. They range between 0.48 and 1.34, but concentrate around 0.9.

The three smallest b-values (shaded in Fig.7) correspond to the regions of East-Upper Austria (b = 0.48), the Vienna Basin (b = 0.63) and Northern Carinthia ($b = 0.64$). The first region is known for shallow earthquakes, and the other two regions have extensional character. They delimit the seismically active Mur-Mürz strike-slip fault system ($b = 0.86$) to the East and to the South.

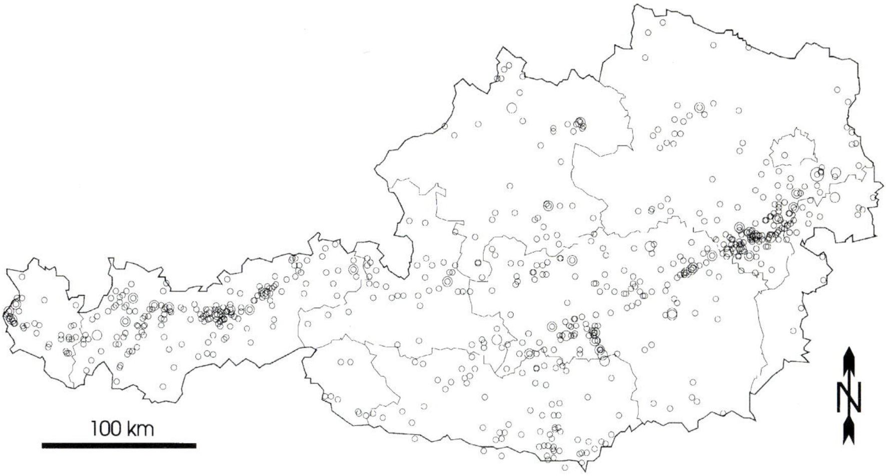

Fig. 5. Distribution of felt earthquakes in Austria between 1900 and 1994.

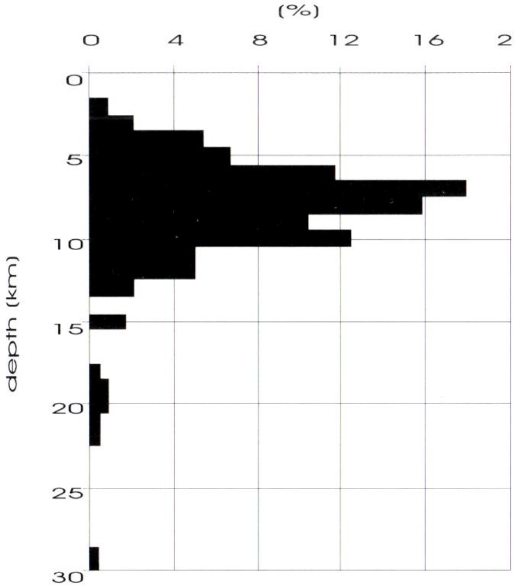

Fig. 6. Focal-depth distribution of earthquakes exceeding a magnitude of 2.5 in Austria between 1900 and 1994 (total number of determined focal depths = 242).

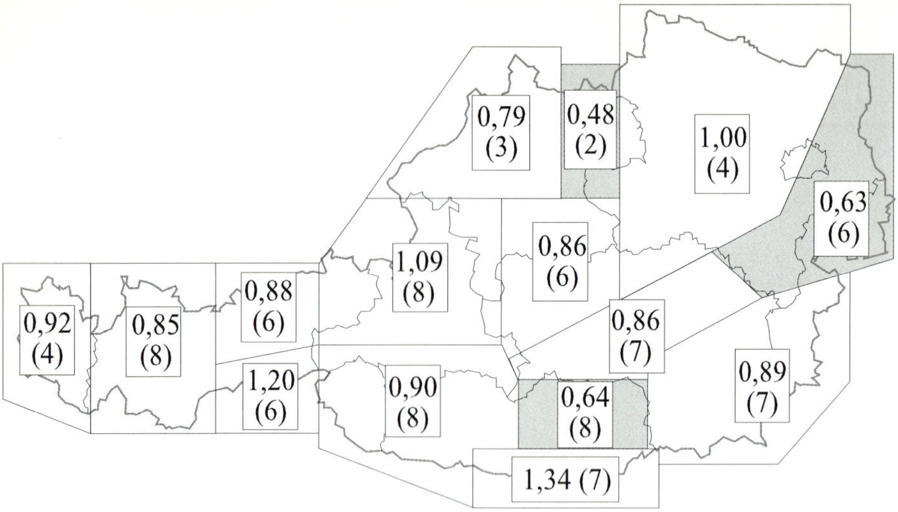

Fig. 7. Regional b-values and average focal depths (in brackets) in Austria.

As already mentioned, the corresponding regional fractal dimensions should amount to twice the b-value (eq.26) of each region. In Figure 7, these values have been omitted on purpose, because D would be based solely on the b-value, - and the relation between D and the b-value can depend on other factors, too.

3.3 Interpretation

Focal depth With few exceptions, most seismic regions in Austria exhibit b-values near unity, although their seismic hazard differs considerably (Lenhardt, 1995) due to the presence of a more or less seismotectonic active fault system.

But we can interpret the results from Figure 7 also in terms of strain instead of seismic hazard. King (1983) pointed out that two-dimensional deformation leads to smaller b-values (theoretically 0.75) and three-dimensional deformation processes result in a b-value of 1.0 - if the b-value is related to the fractal dimension via Aki's formula (eq.26). The three areas in Austria, which exhibit low b-values could therefore be subjected to two-dimensional rather than three-dimensional fault processes. Two-dimensional faulting is equivalent to a plane strain condition in a continuum (King, 1983). Assuming that plane strain conditions exist at a certain depth, this seismotectonic faulting process should originate at a depth which is somewhat different to focal depths of earthquakes from regions subjected to three-dimensional deformation (b-values ~ 1.0). This conclusion can hardly be substantiated by focal depth's statistics, which clearly indicate little differences of average focal depths between the various regions (Fig.7). But another trend becomes apparent when plotting the average focal depth $z_{average}$ of each region against the regional b-value. We recognise an upper bound of the average focal depth as function of the regional b-value, above

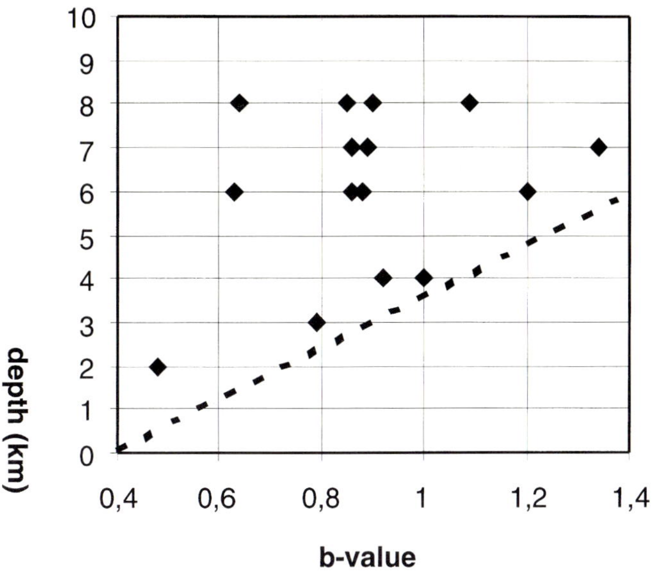

Fig. 8. Regional average of the focal depth versus b-values,

which fewer earthquakes tend to occur. This upper bound (dotted line in Fig.8) can be expressed by

$$z_{\text{average}} > 6 * b - 2.4 \text{ (valid for 1 km} < \text{z} < 6 \text{ km)} \tag{32}$$

with a standard deviation generally less than $z_{\text{average}} - 1$ and concentrating between 2 and 3 km's.

Of course, in all regions shallow earthquakes still occur - but obviously with decreasing frequency, the higher the b-value becomes. In addition, condition (eq.32) could be seen as an expression of the state of stress. The deeper earthquakes occur in the earth's crust, the higher the vertical stress, - and horizontal stresses. Assuming, that earthquake processes occur in a certain stress environment, then the spread of average focal depths in regions of low b-values could also been seen as an expression of similar ratios of horizontal to vertical stresses. Under normal circumstances, a consistent ratio between horizontal to vertical stress is only reached below 3 km of depth (see also Brady and Brown, 1985). In regions of lower b-values, this consistent ratio seems to be reached already at very shallow depth - implying, that horizontal stresses - which normally exceed the vertical stress at shallow depth by far - are much smaller than normal at shallow depth (< 3 km).

As we move along the depicted slope in Figure 8, we could argue according to King (1983), that we are leaving two-dimensional faulting - and enter a regime, which is dominated by three-dimensional deformation processes. This interpretation is of course only valid, as long as we accept Aki's definition of the fractal dimension D, which hinges on the barrier model. The latter represents a self-similar fault model on its own. On the contrary, deeper earthquakes - e.g. below a focal depth of 6 km - do also occur in regions of low b-values. We can attribute this effect either to prevailing three-dimensional deformation (above the slope in Fig.8), - or we have to expect different tectonic mechanisms. The latter would be supported by the low b-values in the Vienna Basin ($b = 0.63$) and Northern Carinthia ($b = 0.64$), which are both known for their extensional character - thus implying low horizontal stresses at shallow depth again.

Multi-affinity? Another explanation involves Hirata's *correlation dimension* D_H (eq.30) which was derived from 2D-data. We have seen, that we cannot derive the 3D-equivalent from D_H, but we can determine the remainder R to the next higher Euclidean dimension E - which is in this case '3' (Tab. 4) - of the 2D-based correlation dimension D_H:

$$R = E - D_H \tag{33}$$

Obviously, the R-factor increases with depth. Now, we are left to speculate. R includes not only the missing part for the 3D-equivalent of the fractal dimension, but also possible self-affine influences, which are expressed by the κ-factor (eq.8). Hence, an increase of R with depth could also mean, that one

Table 4. Possible departure from self-similarity.

z (km)	b-value (see Fig. 8 and eq. 32)	D_H (see eq. 30)	R = 3-D_H
0	0.4	2	1
1	0.6	1.9	1.1
2	0.7	1.8	1.2
3	0.9	1.6	1.4
4	1.1	1.5	1.5
5	1.2	1.4	1.6
6	1.4	1.3	1.7

has to reduce the depth-scale, when comparing focal depths with epicentre-distributions. If (eq.30) and (eq.32) remain valid, then horizontal- and vertical scales of intra-plate earthquakes are possibly not self-similar, not even self-affine, but multi-affine. Multi-affine processes (see also Barabasi and Stanley, 1995) are also affected by other parameters, and do not exhibit a constant κ, and hence R, across a wide range of one scale, - in this case the depth-scale. Nevertheless, the change of R with depth can serve as an indication, that driving mechanisms of earthquakes are not only dependent on depth, but also, that the degree of dependence is depth-dependent again. For vertical stress increases only linearly with depth, other factors should influence the stability of fault-planes. Temperature for instance, or abnormal horizontal tectonic stresses. Here again, the horizontal stresses could be the reason for the observed low *b*-values.

Whatever the case may be, it might be worthwhile to compare the *b*-value of several seismic regions in Austria with the fractal dimension of the corresponding epicentres.

The *b*-value versus the fractal dimension *D* in Austria For calculating the fractal dimension D of the distribution of epicentres in Austria, we can apply the dot-counting method. Although is was already shown, that this method is less reliable (see Tab.3, the box-counting-method), it still constitutes the only kind of calculating the *true spatial* distribution of epicentres, whereas Hirata (1989) calculated D_H from the frequency of distances between *successive* earthquakes. The temporal sequence of epicentres should have been discarded in this regard, because magnitude-statistics (*b*-value) don't consider the sequence of earthquakes either.

The fractal dimension D of epicentres was derived from a grid, in which each cell was allowed to be occupied at least once by an epicentre. The size of each cell was decreased by the square root of 2 for each new calculation-step. The largest cell considered an area of 100 km $*$ 100 km. The smallest cell - or step-size - covered an area of 6.25 km $*$ 6.25 km according to the lowest resolution of epicentral coordinates (appr. 5 km) in the earthquake catalogue. The second column in Table 5 lists the fractal dimension of each polygon, which defines the seismic region. Theoretically this value should be '2.00', because each polygon defines a plane. Applying the dot-counting method, we end up with a mean, which is on average 5 % higher ('2.09' instead of '2.00') with a standard deviation of the same order ('0.11'). Hence, fractal dimensions derived with the dot-counting method tend to overestimate the *true* fractal dimension - very much similar to the box-counting method.

The third column in Table 5 displays the desired fractal dimension of epicentres. We find, that the fractal dimension of the distribution of epicentres varies around 1.25 with a standard deviation of 0.33. This D-value reminds one to the western coastline of Great Britain, or the borderline of Austria. It also resembles the fractal-value of most natural processes again. The question, whether the $D_{\text{Epicentre}}$-values need to be normalized against the fractal dimension D_{Polygon} - which deviates slightly from the theoretical value of '2.00' of the corresponding polygon, is left open for discussion.

The fourth column lists again the *b*-values (see also Fig.7) for reasons of comparison. Averaging the *b*-values of all seismic regions leads us to a mean of 0.9, and a standard deviation of 0.21. To verify whether a correlation between the two parameters exists, we can now compare the fractal dimensions D of epicentre-distributions and their corresponding *b*-values. According to Aki (1981), we should arrive at a positive correlation of the kind $D = 2 * b$, and following Hirata (1989) we should find a negative correlation. As it can been seen from Figure 9, the *b*-value does not seem to correlate with the fractal dimension

Table 5. Fractal dimensions and b-values of seismic regions in Austria (see also Fig. 7).

region	D_{Polygon}	$D_{\text{Epicentre}}$	b-value
1	2.00	0.69	0.79
2	2.19	0.78	0.48
3	2.04	1.17	1.00
4	1.88	1.29	0.63
5	2.05	1.21	0.92
6	2.17	1.68	0.85
7	2.23	1.78	0.88
8	2.04	1.40	1.09
9	2.04	1.34	0.97
10	2.10	1.73	0.86
11	2.13	0.90	1.20
12	1.98	1.27	0.90
13	2.36	1.37	0.64
14	2.05	0.82	0.89
15	2.08	1.38	1.34
mean	2.09	1.25	0.90
std. dev.	0.11	0.33	0.21

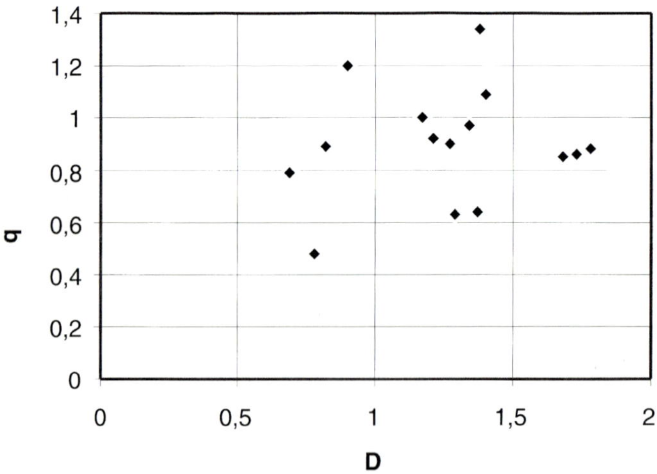

Fig. 9. b-value versus the fractal dimension of epicentres.

(correlation coefficient = + 0.17), even when excluding outliers (correlation coefficient = - 0.34). Therefore, a correlation between the two parameters is left to the reader to avoid further citation. Consequently, the b-value cannot be inferred from the fractal dimension of epicentres - and corrections in terms of dimensions (from 2D to 3D) as well as further discussions regarding multi-affinity become meaningless.

Finally, we should keep in mind, that a fractal dimension of epicentres represents a spatial distribution, whereas the b-value reflects the frequency of magnitudes, hence energies. A comparison of these two parameters seems to be very problematic already from this perspective alone.

3.4 Discussion

Thus, a number of possible causes that lead to abnormal b-values can be quoted: Low b-values (< 1) of a defined region can be a result of

1. horizontal tectonic stresses lower than normal at shallow depth? $\Rightarrow$ earthquakes are less dependent on depth (Fig.8)
2. higher stress drops $\Rightarrow$ likely in an intra-plate scenario (Scholz, 1982)
3. two-dimensional deformation $\Rightarrow$ plane strain (King, 1983)
4. larger correlation dimension? (Hirata, 1989)
5. any combination of (1) – (4),

whereas high b-values (> 1) can be a result of

1. lower stress drops $\Rightarrow$ likely in inter-plate scenarios (Scholz, 1982)
2. fault width independent of thickness of seismogenic layer $\rightarrow$ large earth quakes $M > 6,6$, (Shimazaki, 1986)
3. heterogeneous state of stress $\Rightarrow$ weak crust incapable of sustaining high strains, thus causing earthquake-swarms (Scholz, 1982)
4. associated faulting? $\Rightarrow$ faulting process claiming a volume (Aki, 1981)
5. smaller *correlation dimension*? (Hirata, 1989)
6. any combination of (1) - (5)

Recalling the b-values of Austria (Fig.7), we are tempted now to interpret regions of low and high b-values in terms of the above mentioned points. Firstly, variations of stress drops could explain the observed anomalies. Shallow earthquakes resulting in lower regional b-values might be associated with higher than normal stress drops. The evaluation of seismic records will enable us to verify this hypothesis in future.

Another explanation would involve a larger *correlation* dimension D_H of Hirata (1989). A higher D_H (lower b-value) would then be a result of an activation of a number of faults, which are likely to be 'isolated' - and not associated with each other - in terms of stress transfer. This scenario would not only apply to the two extensional regions, which can be also understood as a tectonic environment where low horizontal stresses - at least in one direction - prevail, but also apply to the other low b-value region in the Bohemian Massif in Upper Austria.

Hence, not only the known extensional character of all three low b-value-regions points towards low horizontal stresses at shallow depth, but also the relation between average focal depth and the b-value (eq.32). In addition, the upper bound (eq.32) could be an indication, that - as the depth increases - stress drops approach standard values (eq.29). The observation, that fewer earthquakes tend to occur at shallow depth on average in regions of high b-values can either be explained by a competent top layer in the crust - or by a relative high cohesion associated with these faults, thus preventing these shallow faults to become seismotectonically active. According to the foregoing discussion, the few earthquakes still originating in this competent top layer of the earth's crust must necessarily utilise existing faults. Lower stress drops could add to this, although they are unlikely in intra-plate scenarios.

Interpretations in terms of the fractal dimension of the spatial distribution of epicentres are not possible because a correlation between this true fractal dimension and the b-value could not be established.

Finally, we can conclude, that deformation processes of all regions, whether they are associated with high b-values or with low b-values, can still be explained by three-dimensional deformation processes.

4 Summary

In earth sciences, the consideration of scaling relations can be useful for

- survey,
- geological,
- geomechanical and
- geophysical

applications, where the optimisation problem '*need of data*' versus '*effort to generate data*' always prevails.

As an example, the borderline of Austria was used for comparing several methods and techniques to calculate fractal dimensions. The results varied strongly around 1.18 - depending on the employed technique, thus making an interpretation extremely difficult. One can certainly argue that the investigated range of scales was too small to tackle the borderline of Austria. However, fractal evaluations stated in literature dealt with a similar degree of resolution (see Mandelbrot, 1982). Hence, the determination of fractal dimensions is very sensitive to employed methods and techniques. One should therefore refrain from comparing results from own calculations with those from literature - unless identical procedures and definitions were used. For example, using the box-counting method led to higher fractal dimensions, when compared with the yard-stick method - similar to the results derived with dot-counting method.

Applying fractal concepts to earthquakes in Austria uncovered a number of interpretations explaining spatial differences in the b-value. Considering the many factors involved for coming up with the relation $D = 2 * b$, it seems that spatial variations of b-values can still be explained by three-dimensional deformation - if we vary the c-value of Hanks and Kanamori's relation (eq.18), thus implying variations in stress drops. Hence, the three areas in Austria, which were identified as regions of having abnormal low b-values, are not necessarily subjected to plane strain conditions, but rather to earthquakes with either higher stress drops or they originate on fault systems in a tectonic environment of low horizontal stresses. The latter explanation seems to be more appropriate.

Interpreting b-values in terms of fractal dimensions, by relating the fractal dimension D strictly to the b-value, can therefore be misleading, for additional factors need to be considered. Stress drops, fault geometry's and the overall state of tectonic stress can cause sufficient deviations from the widely accepted theoretical relation $D = 2 * b$. This fact could serve as an indication, why the fractal dimension D of epicentres was found to be rather independent from the b-value in an intra-plate scenario such as Austria.

Both, the conversion of the b-value to the fractal dimension D - and the interpretation of D, appear to be very controversial. Assuming, that the b-value has been determined correctly, we note grave difficulties to arrive at a fractal dimension. In addition, conclusions based on the fractal dimension can be blurred by preconceptions again. These draw-backs should be kept in mind, when applying fractal concepts.

Because there is no way around the problem of deducing higher order information's from lower order data-sets, we are necessarily left with introducing premises. All premises should be clarified, however - not only when embarking on fractal concepts. Having finally arrived at the desired fractal dimension, we are faced with the problem, that there is more room left for interpretation than for explanation. Nevertheless, the fractal approach not only enables seismologists to arrive at interesting conclusions, but invites them, to investigate the completeness of their data and the various propositions involved.

References

Aki, K., 1981. A probabilistic synthesis of precursory phenomena. In Earthquake Prediction, an International Review, M.Ewing Ser.4, Simpson and Richards (eds.), AGU, 566-574.

Arab, N., A. Kazi and H.H. Rieke, 1994. Fractal geometry of faults in relation to the 12 October 1992 Cairo earthquake. Natural Hazards 10, 221-233.

Barabasi, A.-L. and H.E. Stanley, 1995. Fractal concepts in surface growth. Cambridge University Press.

Brady, B.H.G. and E.T. Brown, 1985. Rock mechanics for underground mining. Allen & Unwin, London.

Brune, J., 1970, 1971. Tectonic stress and the spectra of seismic shear waves from earthquakes. J.Geophys.Res. 75, 1970, 4997-5009 (correction in J.Geophys.Res. 76, 1971, 5002).

Bullen, K.E. and B.A. Bolt, 1987. An introduction to the theory of seismology. Reprint of 4th edition, Cambridge University Press, U.K.

Gutenberg, B. and C.F. Richter, 1954. Seismicity of the Earth and Associated Phenomenon. 2nd edition, University Press, Princeton.

Gutenberg, B. and C.F. Richter, 1956. Magnitude and energy of earthquakes. Ann.Geofis. 9, 1-15.

Grünthal, G., 1993. European Macroseismic Scale 1992 (up-dated MSK-scale). Conseil de l'Europe, Cahiers du Centre Européen de Géodynamique et de Séismologie, Volume 7.

Hanks, T.C. and H. Kanamori, 1979. A moment-magnitude scale. J.Geophys.Res. 84, 2348-2350. Hirata, T., 1989. A correlation between the b-value and the fractal dimension of earthquakes. J.Geophys.Res. 94, 7507-7514.

Huang, J. and D.L. Turcotte, 1990. Are earthquakes an example of deterministic chaos? Geophys.Res.Lett. 17, 223-226.

Hurst, H.E., 1951. Long-term storage capacity of reservoirs. Tr. of the Amer. Soc. of Civil Eng. 116, 770-577 (see also Lomnitz, 1994).

Kanamori, H., 1977. The energy release in great earthquakes. J.Geoph.Res. 82, 2981-2987.

Kanamori, H. and D.L. Anderson, 1975. Theoretical basis of some empirical relations in seismology. Bull.Seism.Soc.Am. 65, No.5, 1073-1095.

Kaye, B., 1993. Chaos & complexity: discovering the surprising patterns of science and technology. VCH Verlagsgesellschaft mbH, Weinheim, Germany.

King, G., 1983. The accomodation of large strains in the upper lithosphere of the earth and other solids by self-similar fault systems: The geometrical origin of b-value. Pure Appl. Geophysics 121, 761-815.

Kolmogorov, A.N., 1941. Über das logarithmische normale Verteilungsgesetz der Dimensionen der Teilchen bei Zerstückelung. Dokl.Akad.Nauk SSSR 31, 99-101 (see Lomnitz, 1994).

Lay, T. and T.C. Wallace, 1995. Modern global seismology. Academic Press, Inc.

Lenhardt, W.A., 1995. Regional earthquake hazard in Austria. Proc. of '10th European Conference on Earthquake Engineering' (Duma, ed.), Balkema, 63-68.

Lomnitz, C., 1994. Fundamentals of earthquake prediction. John Wiley & Sons, Inc.

Mandelbrot, B.B., 1967. How long is the coast of Great Britain? Statistical self-similarity and fractional dimension. Science 156, 636-638.

Mandelbrot, B.B., 1975. Stochastic models for the earth's relief, the shape and the fractal dimension of the coastlines, and the number-area rule for islands. Proc.Nat.Acad.Sci. USA 72, 3825-3828.

Mandelbrot, B.B., 1982. The fractal geometry of nature. Freeman, San Francisco.

Odling, N.E., 1994. Natural fracture profiles, fractal dimensions and joint roughness coefficients. Rock Mech. Rock Engng. 27(3), 135-153.

ÖSZ, 1993. Statistisches Jahrbuch der Republik Österreich 1993. Österreichisches Statistisches Zentralamt, Vienna, Austria.

Richardson, L.F., 1961. The problem of contiguity: An appendix of statistics of deadly quarrels. General Systems Yearbook 6, 139-187 (see also Mandelbrot, 1982).

Richter, C.F., 1958. Elementary Seismology. Freeman & Company. Ryder, J.A., 1988. Excess shear stress in the assessment of geological hazardous situations. J.S.Afr.Inst.Min.& Metall. 88, No.1, 27-39.

Scholz, C.H., 1982. Scaling laws for large earthquakes: consequences for physical models. Bull.Seism.Soc.Am. 72, 1, 1-14.

Seidel, J.P. and C.M. Haberfield, 1995. Towards an understanding of joint roughness. Rock Mech. Rock Engng. 28(2), 69-92.

Shimazaki, K., 1986. Small and large earthquakes: The effects of the thickness of the seismogenic layer and the free surface. In 'Earthquake Source Mechanics', Das, Boatwright and Scholz (eds.), American Geophysical Union, Geophys.Mono.37 Washington, D.C., 209-216 (see also Lay and Wallace, 1995).

Sornette, A., Davy, Ph. and Sornette, D. 1993. Fault growth in brittle-ductile experiments and the mechanics of continental collisions. J.Geoph.Res., Vol.98, No.B7, 12111-12139.

Spottiswoode, S.M. and A. McGarr, 1975. Source parameters of tremors in a deep-level gold mine. Bull.Seism.Soc.Am. 65, 93-112.

Turcotte, D.L., 1986. Fractals and fragmentation. J.Geoph.Res., Vol.91, No.B2, 1921-1926.

Turcotte, D.L., 1989. A fractal approach to probabilistic seismic hazard assessment. Tectonophysics 167, 171-177.

Turcotte, D.L., 1992. Fractals and chaos in geology and geophysics. Cambridge University Press.

Utsu, T., 1965. A method for determining the value of b in a formula log n = a - b * M showing the magnitude-frequency relation for earthquakes. Geophys.Bull.Hokkaido Univ. 13, 99-103 (in Japanese). Referred to in Hirata (1989).

Zhang, Y., Y.P. Chugh and A. Kumar, 1995. Reconstruction and visualization of 3D fracture network using random fractal generators. 'Mechanics of jointed and faulted rock II', (Rossmanith, ed.), Balkema, 543-548.

A plasticity model for discontinua

Yves M. Leroy[1] and William Sassi[2]

[1] Laboratoire de Mécanique des Solides, Ecole Polytechnique, U.M.R. C.N.R.S. no. 7649 91128 Palaiseau Cedex, France
[2] Institut Français du Pétrole, BP 311, 92506 Rueil-Malmaison, France

Abstract. This article is concerned with the development and application of a simple continuum theory for rocks that may contain both randomly as well as preferentially oriented plane discontinuity surfaces. The theory stipulates that displacement discontinuities are independently activated on these surfaces as soon as an appropriate yield criterion is fulfilled; these displacement jumps account for the irreversible, 'plastic' part of the bulk deformation. In stress space, the critical conditions for the activation of discontinuous slip or opening displacements define an overall yield envelope that could be initially anisotropic, reflecting for example a weakness of certain orientations due to pre-existing joint sets. For the yield conditions studied in this paper, essentially a Coulomb-type friction law and a simple fracture opening condition, the inferred stress-strain response under typical triaxial loading conditions reveals the sensitivity of the two discontinuous deformation modes to the confining pressure. The incipient growth of a geological fold in such a material is modelled as a problem of plate bending. The slip- and opening-modes of deformation are found to be activated typically in the fold intrados and extrados, respectively. Under certain conditions, both modes will be activated simultaneously at the same locality and contribute to the total deformation. Field observations on a well exposed sandstone anticline are reported here, which support this conclusion. The present plasticity model for discontinua can clearly be explored in more detail for realistic distributions of faults and joints taken from field observations. It could also be improved in various ways in its description of the underlying deformation mechanisms. Apart from its interest as a mechanical constitutive model, it can also serve as a point of departure for studies of stress-sensitive, anisotropic permeability distributions in fractured formations.

1 Introduction

Natural fracture systems can provide a record of the history of stress, pore fluid pressure, or tectonic deformation, and an understanding of their genesis can explain and help to predict large-scale permeability trends in certain hydrocarbon reservoirs. Mechanical models that relate properties of natural fracture systems, such as fracture orientations or densities, to the tectonic stress and deformation history are therefore of fundamental interest in this context. It is hoped that the simple elasto-plasticity law presented in the following may illustrate the potential of continuum theory to deal with this problem.

Various classifications of natural fractures in terms of their origin have been proposed in the literature (see, e.g., Nelson, 1985; Price & Cosgrove, 1990). A simple scheme distinguishes between tectonic fractures and regional or systematic joints. The former owe their name to the fact that they tend to accommodate tectonic deformation while being generated. Their growth results in complex 3-D geometries which often reveal an interaction of neighboring discontinuities. Lithological layering with sharp rheological contrasts and slip along bedding planes is also a key to the understanding of these tectonic fracture patterns. The origin of systematic joints, the second class of fractures, is still debated. These fractures are very regular in orientation and may cover vast areas. They are formed without significant tectonic deformation although their orientation is often considered as a marker of the paleo-stress state (Rawnsley *et al.*, 1992; Dunne & Hancock, 1993; Petit *et al.*, this Volume). They may also participate in tectonic deformation postdating their genesis. An ideal predictive mechanical model should include the salient features of these discontinuities of different type and provide estimates of fracture densities and orientation based upon a knowledge of the overall deformation history.

In this paper, the link between pervasive fractures and permanent deformation is explored theoretically by means of a simple elasto-plasticity model in which the permanent deformation results from discontinuous (opening or slip) displacements along continuously distributed planes of weakness. The model represents an application to pervasively faulted rocks of ideas that may be traced back at least to the early work of Batdorf & Budiansky (1949) on the properties of polycrystalline materials that deform by crystal-plastic slip within single crystals. Certain of its features are also reminiscent of Reches' (1983) three-dimensional model of faulted rock. It is assumed that there exists an elementary volume of the rock mass under consideration, for which a macro-scale constitutive model can be meaningfully developed. This volume element contains an infinite number of potential discontinuity surfaces with random orientation prior to deformation. Phenomenological laws are introduced that specify the resistance of a discontinuity to slip and opening: A Coulomb friction law with cohesive hardening, to model resistance to slip, and an independent resistance law for irreversible opening displacements, which acts as a cut-off for Coulomb frictional behaviour at low confining pressures and will allow opening to occur instead of or concurrently with slip. The phenomenological laws that characterize the response of a discontinuity could in fact represent distinct physical micromechanisms, including fracture nucleation and growth behaviour. Rather than giving an explicit treatment, these various effects are lumped together in the present simplified phenomenological description with the uniform elastic macro-scale response of the surroundings to localized slip or opening displacements. The micromechanical parameters of the model could exhibit a dependence on the orientation of the discontinuities, such that the overall resistance of the rock mass to deformation would be weaker, in a sense to be defined later, in certain directions, reflecting the presence of pre-existing systematic joint sets. However, in all cases the density of these discontinuities in a representative volume element is such that the macro-scale deformation will be continuous in space.

The idea to construct the macro-scale deformation by the superposition of micro-scale contributions from discontinuous opening or slip displacements along fractures is not new and has been pursued previously in various branches of mechanics including mechanical and civil engineering, geotechnics and geology. For this reason, the short review which follows is certainly far from complete. Oertel (1965) suggested that the deformation observed on his soft clay models is accommodated mainly by slip along four families having orthorhombic symmetry of parallel, closely spaced surfaces. The model of damage in concrete and rocks under tensile load proposed by Bažant & Oh (1985) differs from Batdorf and Budiansky's theory, apart from the choice of micromechanism, mainly by the proposition of a uniform strain over the representative volume element. The tunnel and slope stability analysis of Zienkiewicz & Pande (1977) relies on the idea that the failure of rock masses is due to slip and opening of inherited fracture sets that act as planes of weakness. The tectonic origin of fracture sets and their orientation with respect to the principal stress directions was addressed by Wallace (1951) & Bott (1959). A plane of fracture initiates from a heterogeneity which is then described by a micromechanism such as a penny-shape crack along which frictional sliding is accommodated (Kachanov, 1982a). Local tensile stresses can result in the propagation of branched cracks which is often the micromechanism invoked to model rock dilatancy (Kachanov, 1982b). The macroscopic stress-strain relations based on such mechanisms are derived using, for example, a self-consistent scheme (Horii & Nemat-Nasser, 1983). Sliding along randomly oriented microcracks is often the only relevant mechanisms considered for rocks under overall compressive loading. The angular range of activated cracks depends then only partly on the history or loading path (Lehner & Kachanov, 1995), as was already suggested by Sanders (1954), who employed the plasticity formalism of multiple loading functions of Koiter (1953). Phenomenological plasticity models with the features suggesting the presence of sliding microcracks have been proposed in the past (Rudnicki & Rice, 1975), emphasizing the importance of the micromechanism of deformation on the conditions for overall rock failure in a shear-band mode. A similar influence on the critical tectonic stresses required for the initiation of folds is discussed in detail by Leroy & Triantafyllidis (this Volume).

The contents of this paper are as follows. The next section contains a small-strain formulation of the plasticity model, applicable to the modeling of tectonic deformation in three dimensions. This model could easily be extended so as to allow for large deformations. The third section is devoted to a detailed analysis of a triaxial test in both extension and compression. In the latter case, a plane of weakness is also accounted for in a manner which is similar to the proposition of Jaeger (1960). While only slip can be generated in the compressive triaxial test, both opening and slip modes are activated during extension under triaxial-stress conditions. This distinction is important for a proper understanding of the initial plastic flow in a developing fold, which is studied in the fourth section as a problem of plate bending; extension generated in the extrados of the bent plate is found to result in the opening of discontinuities. In the concluding discussion, the scope of these theoretical results as an aid in the prediction and interpretation and natural fracture systems is critically evaluated for a well-exposed, densely fractured anticline of Devonian sandstone in South Morocco (Gaulier *et al.*, 1996).

2 The plasticity model

The constitutive relations developed in this section pertain to the mechanical response of an elementary volume of a rock mass. The tractions acting on that volume are taken to be those generated by a uniform 'macrostress' field. The inelastic or 'plastic' deformation behaviour of the element at sufficiently high stress levels is assumed to exhibit a strong dependence on the mean stress, or pressure. It is this pressure sensitivity, which is best known from dry friction behaviour of sliding surfaces, that has suggested to us the idea of a plasticity model in which the actual physical micromechanisms of deformation (i.e., grain boundary sliding, growth of transgranular cracks followed by grain crushing, crack growth and interaction on the scale of many grains, etc.) appear lumped together on a continuum scale, where they give rise to a behaviour that can be represented mathematically by the concept of multiple yield-surfaces, as in the work of Koiter (1953). In such a model material, each yield surface represents the constraint imposed on the state of stress by the strength behaviour of real or imagined *plane surfaces of discontinuity* in the displacement[1]. Despite the somewhat ficticious nature of these pervasive discontinuities, it will be helpful to imagine a volume of rock that contains a pervasive system of potential discontinuity surfaces of all orientations, potential, because in order to contribute its share to the bulk strain of the elementary volume, any such discontinuity must first be activated. Moreover, when taken as real, these pervasive discontinuities will suggest the right kind of mathematical continuum model for a rock mass that contains discrete sets of regional fractures or systematic joints.

The range of orientations of the potential planes of discontinuity, as defined by their unit normal vectors $\boldsymbol{n}$, extends over the hemisphere H (of unit radius) shown in Figure 1. Accordingly, every unit normal vector $\boldsymbol{n}$ defines a plane in an oriented material sample, which can accommodate jumps in tangential or

[1] The distinction made in this paper between 'microscale' and 'macroscale' quantities is not so much one between quantities defined at different length scales, but one that differentiates between quantities associated with individual discontinuities and averages of such quantities taken over an appropriate ensemble of discontinuities. 'Micro-' and 'macroscale' quantities thus designate different levels of detail in description, rather than different physical length scales and quantities of both kinds remain essentially macroscopic physical variables in this paper. However, a genuine question of (length) scale that arises in the present context is identified, if not analysed, at the end of this section.

normal displacement. It is assumed that the magnitudes of these jumps are not affected by the presence of neighbouring discontinuities. It is further stipulated that any increase in magnitude of these irreversible discontinuous displacements will necessitate an increase in the macrostress, so that the material element exhibits an overall work hardening behaviour. This assumption ensures a stable response on the macroscale. Its absence would limit the load to a maximum, attained at the first activation of a discontinuity, and this would immediately enforce localized faulting on the macroscale. (For further discussion of these issues, see also the article by Leroy & Triantafyllidis in this Volume).

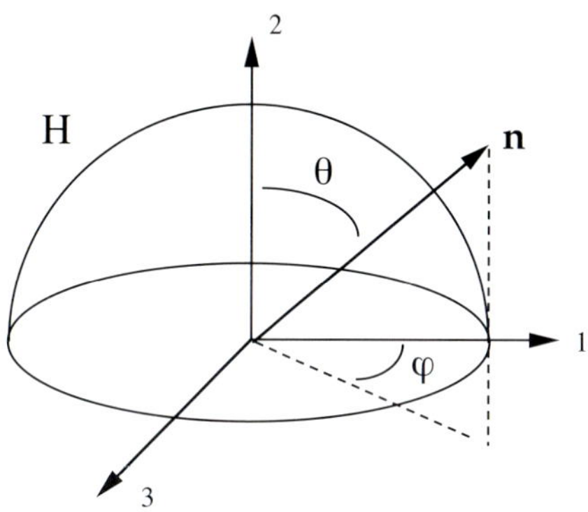

Fig. 1. The hemisphere H defining the range of orientations $\boldsymbol{n}$ of potential planes of discontinuity

The macro-scale strain rate $\dot{\boldsymbol{\epsilon}}$ is composed of an elastic and a plastic part, the former being related to the stress rate by Hooke's law. The plastic part of the strain rate is denoted by $\dot{\boldsymbol{\epsilon}}^p$. Bold quantities identify vectors and second-order tensors. A superposed dot denotes differentiation with respect to time. The rate of macro-scale plastic straining $\dot{\boldsymbol{\epsilon}}^p$ results from the additive contributions $\dot{\boldsymbol{\epsilon}}^p(\boldsymbol{n})$ of all active discontinuities of orientation $\boldsymbol{n}$ on the hemisphere H:

$$\dot{\boldsymbol{\epsilon}}^p = \int_H \dot{\boldsymbol{\epsilon}}^p(\boldsymbol{n})\, dH \,. \tag{1}$$

In order to obtain the permanent or 'plastic' strain rate associated with the discontinuity of normal $\boldsymbol{n}$, it is necessary to identify the force acting on that plane. It is assumed that the local stress vector, $\boldsymbol{t}(\boldsymbol{n})$, acting on the plane may be obtained directly from the macro-scale stress tensor $\boldsymbol{\sigma}$ by the relation [2]

[2] A single dot is used to denote the scalar or 'dot product' of two vectors, the product of a tensor and a vector as in (2) with components $t_i = \sigma_{ij} n_j$, or the product of

$$\boldsymbol{t}(\boldsymbol{n}) = \boldsymbol{\sigma} \cdot \boldsymbol{n} \,. \tag{2}$$

The stress vector may be viewed as composed of a normal and a tangential component according to $\boldsymbol{t}(\boldsymbol{n}) = \boldsymbol{t}_N(\boldsymbol{n}) + \boldsymbol{t}_T(\boldsymbol{n})$, as shown in Figure 2. The normal stress $\sigma_N(\boldsymbol{n})$ on the plane of orientation $\boldsymbol{n}$, i.e., the magnitude the component $\boldsymbol{t}_N(\boldsymbol{n})$ is given by [3,4]

$$\sigma_N(\boldsymbol{n}) = \boldsymbol{n} \cdot \boldsymbol{t}(\boldsymbol{n}) = \boldsymbol{n} \cdot \boldsymbol{\sigma} \cdot \boldsymbol{n} = (\boldsymbol{n} \otimes \boldsymbol{n}) : \boldsymbol{\sigma} \,. \tag{3}$$

Writing $\boldsymbol{t}_N(\boldsymbol{n}) = \sigma_N(\boldsymbol{n})\boldsymbol{n} = (\boldsymbol{n} \cdot \boldsymbol{t}(\boldsymbol{n}))\boldsymbol{n} = (\boldsymbol{n} \otimes \boldsymbol{n}) \cdot \boldsymbol{t}(\boldsymbol{n})$, for the normal component, also yields [5]

$$\boldsymbol{t}_T(\boldsymbol{n}) = \boldsymbol{t}(\boldsymbol{n}) - \boldsymbol{t}_N(\boldsymbol{n}) = (\boldsymbol{I} - \boldsymbol{n} \otimes \boldsymbol{n}) \cdot \boldsymbol{\sigma} \cdot \boldsymbol{n} \,, \tag{4}$$

for the tangential component of the stress vector. This component defines a unit tangent vector $\boldsymbol{m}$ in the plane of orientation $\boldsymbol{n}$ by

$$\boldsymbol{m}(\boldsymbol{n}) = \frac{\boldsymbol{t}_T(\boldsymbol{n})}{\tau(\boldsymbol{n})} \,, \tag{5}$$

in which the resolved shear stress $\tau(\boldsymbol{n})$ is defined as the magnitude of $\boldsymbol{t}_T(\boldsymbol{n})$. This shear stress is conveniently expressed in terms of the macrostress and the unit vectors $\boldsymbol{m}$ and $\boldsymbol{n}$ by

$$\tau(\boldsymbol{n}) = \boldsymbol{m} \cdot \boldsymbol{\sigma} \cdot \boldsymbol{n} = \frac{1}{2}\left[\boldsymbol{m} \otimes \boldsymbol{n} + \boldsymbol{n} \otimes \boldsymbol{m}\right] : \boldsymbol{\sigma} \,, \tag{6}$$

where the symmetry of the stress tensor has been exploited to introduce the symmetric Schmid tensor as a factor forming a scalar product with the stress. Note also that the rate $\dot{\tau}$ is related to the time derivative of the stress tensor $\dot{\boldsymbol{\sigma}}$ in the same way.

---

two second-order tensors $\mathbf{C} = \mathbf{A} \cdot \mathbf{B}$ with components $C_{ij} = A_{ik}B_{kj}$ in a Cartesian coordinate system. Here repeated indices imply summation over the range of the index, i.e., from 1 to 3.

[3] Note that compressive stresses are taken negative throughout this paper.

[4] The dyadic or tensor product $\mathbf{a} \otimes \mathbf{b}$ of two vectors $\mathbf{a}$ and $\mathbf{b}$ constitutes a second-order tensor with components $(\mathbf{a} \otimes \mathbf{b})_{ij} = a_i b_j$; it is defined through its action on an arbitrary vector $\mathbf{c}$ by $(\mathbf{a} \otimes \mathbf{b})\mathbf{c} = (\mathbf{c} \cdot \mathbf{b})\mathbf{a}$. The colon on the right-hand-side stands for the scalar product of any two tensors $\mathbf{A}$ and $\mathbf{B}$, i.e., the scalar $\mathbf{A} : \mathbf{B} = A_{ij}B_{ij}$ in terms of the Cartesian components of the two tensors. When $\mathbf{A}$ or $\mathbf{B}$ is symmetric, the product also equals the 'trace' $\mathrm{tr}(\mathbf{A} \cdot \mathbf{B}) = A_{ij}B_{ji}$ of the product $\mathbf{A} \cdot \mathbf{B}$.

[5] $\mathbf{I}$ is the second-order identity tensor, with components $\delta_{ij} = 1$, if $i = j$, and $\delta_{ij} = 0$ otherwise.

Having identified the forces acting on the discontinuity of normal $\boldsymbol{n}$, we now propose a phenomenological plasticity law to describe the two mechanisms of interest: the generation of discontinuous slip and opening displacements on a plane surface of discontinuity. For slip to occur on a given discontinuity surface, we require that the equality in the Mohr-Coulomb yield criterion

$$\phi^S(\boldsymbol{\sigma}, \boldsymbol{n}, \gamma(\boldsymbol{n})) \equiv \tau(\boldsymbol{n}) + \mu\sigma_N(\boldsymbol{n}) - c_0(\gamma(\boldsymbol{n})) \leq 0\,, \tag{7}$$

be satisfied. Here μ is a constant coefficient of friction and c_0 is the cohesion, which is taken to be a function of the *internal variable* $\gamma(\boldsymbol{n})$. Note that these strength parameters could both be functions of the orientation of the plane of weakness. The internal variable is conceived here as dimensionless equivalent plastic shear strain and is defined by

$$\gamma(\boldsymbol{n}) = \int_0^t \sqrt{2\,\dot{\boldsymbol{\epsilon}}_S^{p\prime}(\boldsymbol{n}) : \dot{\boldsymbol{\epsilon}}_S^{p\prime}(\boldsymbol{n})}\,, \tag{8}$$

where $\boldsymbol{\epsilon}_S^{p\prime}(\boldsymbol{n})$ is the deviator[6] of the plastic strain $\boldsymbol{\epsilon}_S^p(\boldsymbol{n})$ that is generated by slip on discontinuity planes of orientation $\boldsymbol{n}$. The equivalent plastic strain rate in shear, $\dot{\gamma}(\boldsymbol{n})$, for a discontinuity of orientation[7] $\boldsymbol{n}$ may be determined from the so-called 'consistency condition' which calls for the continuing satisfaction of the yield criterion (7) as a condition for continuing plastic straining. The condition therefore demands that $\dot{\phi}^S(\boldsymbol{\sigma}, \boldsymbol{n}, \gamma(\boldsymbol{n})) = 0$. After substitution in (7) of the expressions (3) and (6) for $\sigma_N(\boldsymbol{n})$ and $\tau(\boldsymbol{n})$ in terms of the macrostress, and differentiation, this produces the following result for the plastic rate of shearing:

$$\dot{\gamma}(\boldsymbol{n}) = \frac{1}{h^S}\frac{\partial \phi^S}{\partial \boldsymbol{\sigma}} : \dot{\boldsymbol{\sigma}} \quad \text{with} \quad \frac{\partial \phi^S}{\partial \boldsymbol{\sigma}} = \frac{1}{2}\big[(\boldsymbol{m} + \mu\boldsymbol{n}) \otimes \boldsymbol{n} + \boldsymbol{n} \otimes (\boldsymbol{m} + \mu\boldsymbol{n})\big]\,. \tag{9}$$

Here h^S stands for the derivative of the function $c_0(\gamma)$. Note that $\partial\phi^S/\partial\boldsymbol{\sigma}$ defines the normal to the yield surface ϕ^S in stress space.

We must now identify the plastic deformation rate $\dot{\boldsymbol{\epsilon}}_S^p(\boldsymbol{n})$ which contributes to the microstrain $\boldsymbol{\epsilon}_S^p(\boldsymbol{n})$ whenever the loading conditions $\phi^S = 0$ and $\frac{\partial\phi^S}{\partial\boldsymbol{\sigma}} : \dot{\boldsymbol{\sigma}} > 0$ are fulfilled. It is constructed by assuming that slip on a plane with surface normal $\boldsymbol{n}$ takes place in the direction $\boldsymbol{m}$ of the tangential component of the stress vector and that the deviator of the resulting micro-strain rate is proportional to the Schmid tensor:

[6] The deviator of a tensor $\mathbf{A}$ is defined by $\mathbf{A}' = \mathbf{A} - \frac{1}{3}(\mathbf{A} : \mathbf{I})\mathbf{I}$.

[7] Although we speak here of a discontinuity of orientation $\mathbf{n}$ and an associated plastic strain, it would have been more appropriate to start from the picture of a certain *number density* of parallel discontinuities transsecting the elementary volume and, in general, assume an orientation dependence for this density. In the present continuum treatment, the density parameter may however be thought of as having been absorbed in the nondimensional internal strain variable $\gamma(\mathbf{n})$ through the hardening response specified in (9).

$$\dot{\epsilon}_S^{p\,\prime} = \dot{\gamma}(\boldsymbol{n})\frac{1}{2}[\boldsymbol{m}\otimes\boldsymbol{n} + \boldsymbol{n}\otimes\boldsymbol{m}]\,. \tag{10}$$

At least three interpretations can be given to this expression, interpreted as the micro-scale contribution to the macro-scale strain rate introduced by (1). Thus, it is first observed that the right-hand side of (1) has a structure that is familiar from theories of crystal plastic deformation (see, e.g., the review by Asaro, 1983) in which the integral is typically replaced by a sum over a finite number of slip systems. The tangential vector $\boldsymbol{m}$ may however be interpreted differently in these theories as being determined by a crystallographic direction rather than the resolved macrostress.

In a second interpretation of (1) and (10), the integration over all orientations is again replaced by a finite sum, this time over preferentially oriented sets of faults in a fractured formation, as in the work of Gauthier & Angelier (1985). The present model differs from theirs mainly in that a constitutive relation is invoked for the displacement along every discontinuity.

Expression (10) also has an interpretation for slip occurring under dynamic conditions. The Schmid factor on the right-hand side of (10) then equals the seismic moment tensor density, m_{qp}, normalized by $2G$, where G is the shear modulus of an isotropic elastic medium (Molnar, 1983). The difference between the present static and the dynamic interpretation of (10) is thus that in the former $\dot{\gamma}$ depends on the rheology of the sliding discontinuities rather than on the elastic properties of the surrounding rock mass.

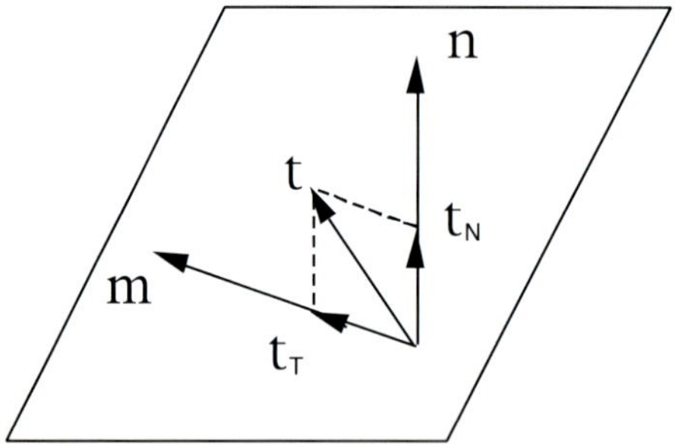

Fig. 2. Decomposition of stress vector into normal and tangential component and identification of potential slip direction **m** for a plane with unit normal **n**.

Dilatancy during slip results from the presence of stiff heterogeneities and asperities on the sliding surfaces. Following Rudnicki & Rice (1975), the volumetric strain rate is therefore defined in terms of the equivalent strain rate in shear and a dilatancy coefficient β that may depend on γ. In the following, this coefficient is assumed to be constant and independent of the orientation of a discontinuity. The microstrain resulting from the dilation of a discontinuity of orientation $\boldsymbol{n}$ is thus taken to be proportional to $\dot{\gamma}(\boldsymbol{n})\boldsymbol{n} \otimes \boldsymbol{n}$, with β as the factor of proportionality. The total strain rate associated with slip on a dilating discontinuity thus becomes

$$\dot{\boldsymbol{\epsilon}}^p_S(\boldsymbol{n}) = \dot{\gamma}(\boldsymbol{n})\big[\frac{1}{2}(\boldsymbol{m} \otimes \boldsymbol{n} + \boldsymbol{n} \otimes \boldsymbol{m}) + \beta \boldsymbol{n} \otimes \boldsymbol{n}\big]\,. \tag{11}$$

and this contains the essence of our plasticity model for slip along a discontinuity of orientation $\boldsymbol{n}$. Note that for the sake of simplicity the dependence of $\dot{\gamma}(\boldsymbol{n})$ on the macro-scale stress state has not been made explicit in the above expressions.

Let us now consider, as a second micro-scale deformation mechanism, the opening of discontinuities independent of any slip displacement. Such a mechanism could already be activated at small compressive normal stresses ($\sigma_N(\boldsymbol{n}) < 0$) and its operation can be allowed for by the introduction of a second yield criterion of the form

$$\phi^O(\boldsymbol{\sigma}, \boldsymbol{n}, d(\boldsymbol{n})) \equiv \sigma_N(\boldsymbol{n}) - k_0(d(\boldsymbol{n})) = 0\,, \tag{12}$$

which involves only the normal stress σ_N and a material function $k_0(d)$ of the accumulated opening strain, d, an internal variable that may depend on the orientation of the discontinuity. This criterion acts as a cut-off of the Mohr-Coulomb criterion in stress space, as is illustrated Figure 3. Note that $k_0(d(\boldsymbol{n}))$ is taken to be initially negative, which excludes the state of zero stress from the initial elastic domain. This choice is consistent with numerous observations of the damage induced in core samples retrieved from great depth. It makes allowance for the operation of one or more micromechanisms, not to be discussed here, that effect a transformation of compressive macrostresses into locally tensile states of stress. k_0 is taken to be an increasing function of d, reflecting an increasing resistance of the discontinuity to opening. The structure of the plastic strain produced by this second deformation mechanism resembles that associated with slip-induced dilation in (11), if $\dot{\gamma}\beta$ is replaced by $\dot{d}$. The rate of plastic deformation associated with this opening mechanism for a discontinuity of orientation $\boldsymbol{n}$ is thus written

$$\dot{\boldsymbol{\epsilon}}^p_O(\boldsymbol{n}) = \dot{d}(\boldsymbol{n})\,\boldsymbol{n} \otimes \boldsymbol{n}\,. \tag{13}$$

Here $\dot{d}$ is determined from the consistency condition $\dot{\phi}^O = 0$ imposed on the yield criterion (12), giving

$$\dot{d}(\boldsymbol{n}) = \frac{1}{h^O}\frac{\partial \phi^O}{\partial \boldsymbol{\sigma}} : \dot{\boldsymbol{\sigma}}\,, \quad \text{with} \quad \frac{\partial \phi^O}{\partial \boldsymbol{\sigma}} = \boldsymbol{n} \otimes \boldsymbol{n}\,, \tag{14}$$

in which the scalar h^O denotes the derivative of the function $k_0(d)$. Note that the opening mechanism is activated, if the conditions $\phi^O = 0$ and $(\partial\phi^O/\partial\boldsymbol{\sigma}) : \dot{\boldsymbol{\sigma}} > 0$ are simultaneously satisfied.

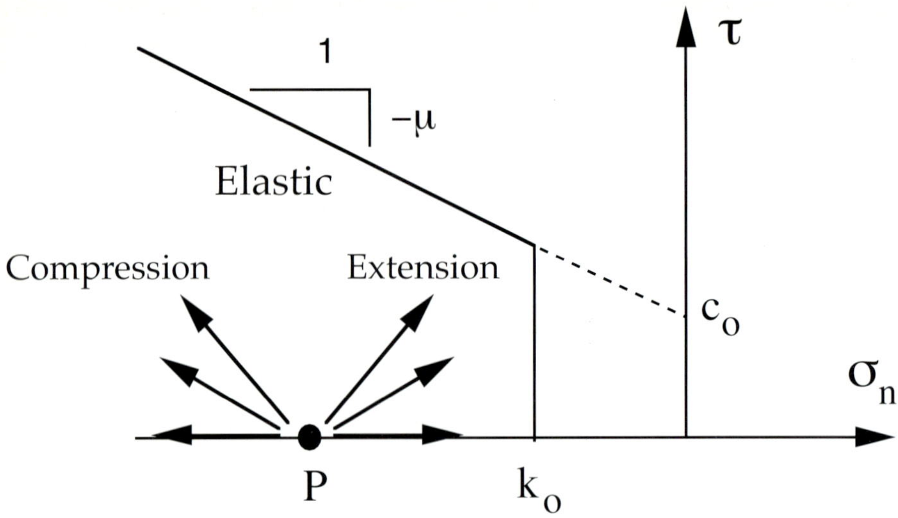

Fig. 3. The yield locus in stress space for a potential discontinuity: A Mohr-Coulomb envelope truncated by the opening criterion $\sigma_N = k_0(d)$.

The second mechanism, which potentially excludes the zero stress point from the admissible domain, may raise a question about the stability of the postulated model material in the sense of Drucker (1951). To dispel such concerns, it suffices to consider a single discontinuity of orientation $\boldsymbol{n}$ and a stress state $\boldsymbol{\sigma}^*$ in the elastic domain or on the loading surface. Identify, for example, $\boldsymbol{\sigma}^*$ with point P in Figure 3. The material is stable in the sense of Drucker, if the work done by an external loading process in a cycle starting at $\boldsymbol{\sigma}^*$ and resulting in plastic deformation is strictly non-negative. Drucker showed that this condition requires that

$$(\boldsymbol{\sigma} - \boldsymbol{\sigma}^*) : \dot{\boldsymbol{\epsilon}}^p_O + \dot{\boldsymbol{\sigma}} : \dot{\boldsymbol{\epsilon}}^p_O > 0\,,$$

in which $\boldsymbol{\sigma}$ lies on the loading surface. The second term on the left-hand-side of this inequality is strictly non-negative if $\dot{d}$ is non-zero, given the normality of the plastic flow (13) and as long as there is positive work hardening ($h^O > 0$). The same property and the convexity of the yield condition (12) entail the positiveness of the first term in the above inequality. The assumed deformation mechanism of the opening of discontinuities is thus stable in the sense of Drucker. It should also be recalled here that the relevance of Drucker's postulate as a sufficient, but not necessary condition for the stability of the first mechanism of frictional slip has been repeatedly discussed in the literature (see, for example, the note by Mandel (1964) and related discussions in the same volume) and will not be addressed here. A further point of concern could be the lack of uniqueness of the normal direction to the loading surface at the intersection, in stress space, of the yield surfaces corresponding to the two distinct criteria (7) and (12) (cf. Fig. 3). However, no such problem arises with the present model in view of the independence of the two postulated deformation mechanisms (Koiter, 1953). An indeterminacy could have resulted here, if such an interaction had been allowed for, e.g. by the introduction of latent hardening as it is appropriate for some crystalline materials (Pan & Rice, 1983).

In summary, the total plastic rate of deformation, due to the activation of a planar discontinuity of orientation $\boldsymbol{n}$ by the two independent micromechanisms, is obtained as the sum of the strain rates (11) and (13):

$$\dot{\boldsymbol{\epsilon}}^p(\boldsymbol{n}) = \dot{\gamma}(\boldsymbol{n})\alpha^S(\boldsymbol{n})\big[\frac{1}{2}(\boldsymbol{m}\otimes\boldsymbol{n}+\boldsymbol{n}\otimes\boldsymbol{m})+\beta\boldsymbol{n}\otimes\boldsymbol{n}\big]+\dot{d}(\boldsymbol{n})\alpha^O(\boldsymbol{n})\boldsymbol{n}\otimes\boldsymbol{n}\,. \quad (15)$$

Here the scalar factors $\alpha^S(\boldsymbol{n})$ and $\alpha^O(\boldsymbol{n})$ take on the values 1 or 0, depending on whether or not the corresponding micromechanism is activated.

Having characterized the plastic deformation accumulated on individual discontinuities by the two independent micro-scale deformation mechanisms, we must now determine the resulting macro-scale plastic deformation $\dot{\boldsymbol{\epsilon}}^p$. This last operation involves the substitution of (15) for the integrand in (1), followed by an appropriate integration over the hemisphere H. Before carrying out such a calculation for the triaxial test and the bending of a plate, we wish however to discuss the incorporation in our plasticity model of weak discontinuities of preferred orientation. These discontinuities are taken to represent inherited systematic joints that were formed during earlier episodes of stressing and deformation and will contribute to either slip or opening. In that sense, they are weaker than any potential discontinuity of the rock mass. Systematic joints are also called regional joints because of their large extent. It is assumed, however, that their contribution to the tectonic deformation occurs by slip or opening along patches that are sufficiently well distributed to justify the assumption that the macro-scale stress field prevails on every discontinuity. The present model is built on the assumption that there exists a finite number N of joint sets of a given precise orientation, which cut through the elementary volume. The relevant components of stress acting on a discontinuity plane are assumed to satisfy a plastic yield criterion, similar to the one proposed for the above continuously distributed potential discontinuities, but with different material parameters. The total macroscopic plastic strain rate is composed of the integral term in (1) and a sum $\Sigma_I^N \dot{\boldsymbol{\epsilon}}_a^p(\boldsymbol{n}_a)$ of terms that represent the contributions to the macrostrain from N individual sets of discontinuities. It is thus given by

$$\begin{aligned}\dot{\boldsymbol{\epsilon}}^p &= \int_H \dot{\boldsymbol{\epsilon}}^p(\boldsymbol{n})\,dH \\ &+\sum_{a=I}^{N}\Big\{\dot{\gamma}_a\alpha_a^S\big[\frac{1}{2}(\boldsymbol{m}_a\otimes\boldsymbol{n}_a+\boldsymbol{n}_a\otimes\boldsymbol{m}_a)+\beta_a\boldsymbol{n}_a\otimes\boldsymbol{n}_a\big]+\dot{d}_a\alpha_a^O\boldsymbol{n}_a\otimes\boldsymbol{n}_a\Big\}, \quad (16)\end{aligned}$$

where the $\dot{\boldsymbol{\epsilon}}_a^p(\boldsymbol{n}_a)$ have been assumed to satisfy an expression of the form (15).

A similar superposition of strains resulting from slip on weak discontinuities and microstrains generated pervasively throughout a rock mass has been considered earlier by Jaeger (1960). An obvious problem arising in this context is that of a scale-dependence of certain results, such as for example the calculated stiffness of the material volume element under consideration. In general, one must expect quantities of this kind to depend on the size of the elementary volume by the mere fact that different volumes will sample different ensembles of systematic joints[8]. In the best of all situations, this size dependence will fade as the volume exceeds a certain minimum size that is still small in comparison with a characteristic overall length of the rock mass of interest. One can then speak meaningfully of a *representative* elementary volume (REV), a qualification we have so far avoided on purpose. Indeed, a challenging task for studies of the present kind remains precisely the determination of typical length scales of REV's in the field.

3 The triaxial test

This section is devoted to an analysis of the permanent deformation of an elasto-plastic material of the type just defined, when a representative sample of this material is subjected to standard triaxial loading conditions. Linear hardening laws are adopted for the cohesion and the opening hardening functions in order to obtain analytical expressions. The triaxial compressive test, in which only sliding discontinuities can be activated, is considered first. Triaxial extension, which permits sliding and opening displacements to occur, is discussed in the second part of this section.

Assume now that the material sample is initially subjected to the isotropic compressive stress $-P\boldsymbol{I}$, where P is a positive scalar pressure. The specimen is subsequently loaded by a monotonically increasing or decreasing axial component of stress, q, in direction 2 (cf. Fig. 4), so that the total stress state becomes

$$\boldsymbol{\sigma} = -P\boldsymbol{I} + q\,\boldsymbol{e}_2 \otimes \boldsymbol{e}_2\,, \tag{17}$$

where q varyies monotonically with time and $\boldsymbol{e}_2$ is the base vector associated with the x_2 coordinate axis.

The components of the unit normal $\boldsymbol{n}$ of a potential discontinuity plane in the Cartesian coordinate system of Figure 1, when expressed in terms of the two Euler angles θ and φ, are

$$\begin{pmatrix} n_1 \\ n_2 \\ n_3 \end{pmatrix} = \begin{pmatrix} \sin\theta\cos\varphi \\ \cos\theta \\ \sin\theta\sin\varphi \end{pmatrix}. \tag{18}$$

[8] In expression (16) this would imply a dependence of the number N of preferentially oriented fracture sets on the size of the elementary volume, i.e., on the normalizing length involved in the definition of the nondimensional displacement jumps γ_a and d_a.

The normal stress $\sigma_N(\boldsymbol{n})$, the resolved shear stress $\tau(\boldsymbol{n})$, and the vector $\boldsymbol{m}(\boldsymbol{n})$ are given by

$$\begin{aligned} \sigma_N(\boldsymbol{n}) &= -P + q\cos^2\theta\,, \\ \tau(\boldsymbol{n}) &= |q|\cos\theta\,\sin\theta\,, \\ \boldsymbol{m}(\boldsymbol{n}) &= \mathrm{Sign}(q)\big(-\cos\theta\,\cos\varphi\,\boldsymbol{e}_1 + \sin\theta\,\boldsymbol{e}_2 - \cos\theta\sin\varphi\,\boldsymbol{e}_3\big)\,. \end{aligned} \tag{19}$$

These expressions are found by using (5) and (6) in (4) and (5), while observing that $\tau = |\boldsymbol{t}_T \cdot \boldsymbol{t}_T|^{1/2}$. Note that the normal stress and the resolved shear stress are independent of the Euler angle φ. A direct consequence of this axial symmetry is that the activation of potential discontinuities always occurs for all values of this angle. However, this symmetry will be carried over to the deformation of the sample only in the absence of any discrete sets of joints of preferred orientation.

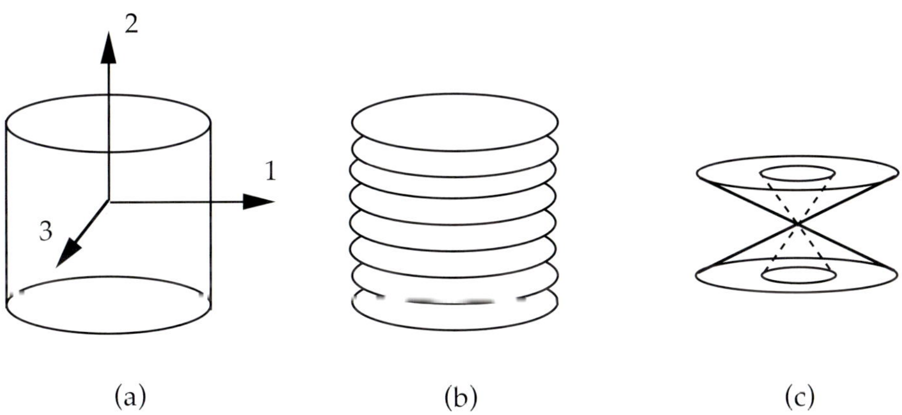

Fig. 4. A triaxial test. The cylidrical specimen (a), initially under uniform pressure, is subjected to an increasing or decreasing axial load. Permanent deformation of the sample results either from opening of horizontal discontinuities (b), or from slip on fault planes whose range of dips is delimited by the tangent planes of the two cones shown in (c).

The first part of this section is concerned with the compressive triaxial test for which the scalar q in (5) is negative and monotonically decreasing. However, the equations to be presented are kept general, allowing q to be of any sign in anticipation of the subsequent analysis of the extension test. Inspection of the evolution of σ_N and τ according to (7) and of the structure of the yield criterion (12) leads to the conclusion that it is impossible to activate the opening mode if q is negative. This conclusion is also evident from the illustration in Figure 3 of the stress path followed during compression. We shall thus focus on the sliding mode in this first part of this section. Inserting the expressions (7) for σ_N and τ in the yield criterion for sliding mode (7) provides the following condition valid throughout the test:

$$\sin(2(\theta + \mathrm{Sign}(q)\phi) \leq [2c_0(\gamma(\boldsymbol{n}))\cos\phi + (2P - q)\sin\phi]/|q|\,, \tag{20}$$

The inequality is satisfied in the elastic range of deformation and we now inspect the conditions for which the equality holds for the first time, defining first yield. Since c_0 increases with γ, we must examine solutions to equation (8) for the initial value $c_0(0)$. The two solutions found are

$$\theta_{1,2} = \theta_c \mp \frac{1}{2}\left[\frac{\pi}{2} - \arcsin\left(\frac{2c_0(0)\cos\phi + (2P - q)\sin\phi}{|q|}\right)\right] \tag{21}$$

where

$$\theta_c = \frac{\pi}{4} - \mathrm{Sign}(q)\frac{\phi}{2} \tag{22}$$

is the classical Coulomb angle, that is the angle of bisection of the first activated pair of discontinuity surfaces by the largest compressive principal stress. Since $0 \leq \phi \leq \pi/2$, this falls into the ranges $\pi/4, (0) \leq \theta_c \leq \pi/2, (\pi/4)$ for $\mathrm{Sign}(q) = -1, (+1)$. The angle $\theta_1 = \theta_2 = \theta_c$ thus represents the solution of (8) obtained for a critical load such that the argument of the arcsin function in (11) drops to the value 1 for the first time, that is for the load

$$q_c = \mathrm{Sign}(q)\frac{2[c_0(0)\cos\phi + P\sin\phi]}{1 + \mathrm{Sign}(q)\sin\phi}\,. \tag{23}$$

Beyond this critical value of q, two solutions, θ_1 and θ_2 will exist for equation (8). These furnish the limits of the orientation range of activated discontinuities $[\theta_1, \theta_2] \equiv \theta_2 - \theta_1$. The existence and growth of this active orientation range has been indicated already in Figure 4c. It is made more precise by Figure 5, where the limits θ_1, θ_2 are displayed in a Mohr diagram, making use of the stress 'pole' at the point $(-P, 0)$ (see, e.g., Mandl, 1988). The fact that the orientation range $[\theta_1, \theta_2]$ grows symmetrically about the Coulomb orientation θ_c is essentially a consequence of the assumed hardening of the slip systems. This permits a continuing increase in the load, which in turn results in the activation of less favourably oriented slip systems and a corresponding growth of the angular range of activated discontinuities. At any given instant, the most critical orientation, θ_c, will have hardened more than any other orientation. The state of stress on that plane is represented by the tangent point of a Coulomb envelope, which has been shifted in accord with the amount of cohesive hardening corresponding to the current value of $\gamma(\theta_c, q)$, and the stress circle passing through the points $(-P, 0)$ and $(-P + q, 0)$. (In constructing Fig. 5, we have made use of (18) and the linear hardening law (17).) Since the stresses on any newly activated planes of orienation θ_1 and θ_2 remain constrained by a Coulomb condition for first slip,

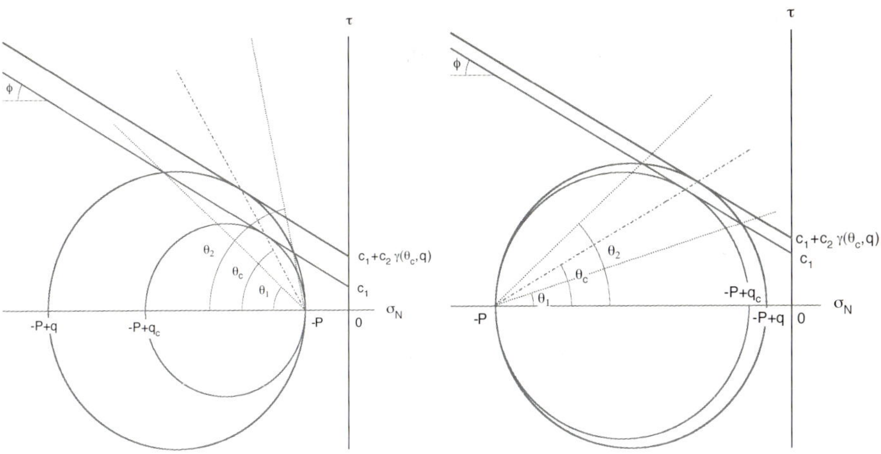

Fig. 5. Graphical solution for the orientation range (11) of activated discontinuities for the parameter values $c_1 = 10\,\mathrm{MPa}$, $c_2 = 50\,\mathrm{MPa}$, and $\mu = \tan\phi = 0.6$. (a) Triaxial compression test at $P = 20\,\mathrm{MPa}$ with $q_c \approx -74.5\,\mathrm{MPa}$. (b) Triaxial extension test at $P = 60\,\mathrm{MPa}$ with $q_c \approx 51.5\,\mathrm{MPa}$.

with $c_0 = c_0(0)$, they must be represented by the points of intersection of the initial Mohr envelope with the current stress circle.

Let us now quantify the shear deformation associated with slip on active discontinuities and from this determine the relation between the applied load and the permanent macroscopic deformation of the sample. If slip-induced dilatancy is disregarded ($\beta = 0$) in addition to opening mode (13) and a correspondingly simplified form of expression (15) is substituted in (16), together with (5),(6), and (7c), there results the following expression for the macro-scale plastic strain rate:

$$\dot{\boldsymbol{\epsilon}}^p = \frac{1}{2\pi}\int_0^{\pi/2}\int_0^{2\pi} \alpha^S(\theta)\dot{\gamma}(\theta)\boldsymbol{R}(\theta,\varphi)\sin\theta\, d\theta d\varphi\,, \tag{24}$$

where $\boldsymbol{R}(\theta,\varphi)$ stands for the Schmid tensor $\frac{1}{2}(\boldsymbol{m}\otimes\boldsymbol{n}+\boldsymbol{n}\otimes\boldsymbol{m})$, the component matrix of which becomes

$$[R_{ij}] = \tag{25}$$

$$\frac{1}{2}\mathrm{Sign}(q)\begin{bmatrix} -\sin 2\theta\cos^2\varphi & (1-2\cos^2\theta)\cos\varphi & -\frac{1}{2}\sin 2\theta\sin 2\varphi \\ (1-2\cos^2\theta)\cos\varphi & \sin 2\theta & (1-2\cos^2\theta)\sin\varphi \\ -\frac{1}{2}\sin 2\theta\sin 2\varphi & (1-2\cos^2\theta)\sin\varphi & -\sin 2\theta\sin^2\varphi \end{bmatrix}.$$

The fact that the strain rate $\dot{\gamma}$ does not depend on the angle φ simplifies the integration of (14). Moreover, on account of (9a) and (5), $\gamma(\theta)$ depends on time only through the time-like factor q; (14) is therefore readily integrated along the loading path to provide the total macroscopic plastic strain:

$$[\epsilon^p_{ij}] = \frac{\mathrm{Sign}(q)}{2}\begin{bmatrix} -1 & 0 & 0 \\ 0 & 2 & 0 \\ 0 & 0 & -1 \end{bmatrix} \int_0^{\pi/2} \alpha^S(\theta)\gamma(\theta)\sin^2\theta\cos\theta d\theta\,. \tag{26}$$

To render explicit the relation between this plastic macrostrain and the applied load, the functional dependence of the cohesion c_0 on the internal variable $\gamma(\boldsymbol{n})$ must be specified for all orientations $\boldsymbol{n}$ of potential discontinuity surfaces. For simplicity, we consider the linear relation

$$c_0(\gamma) = c_1 + c_2\gamma\,. \tag{27}$$

Here c_1 and c_2 are constants that are independent of the orientation $\boldsymbol{n}$. A relation of this type can be motivated for crack-like discontinuities which, in the absence of crack growth, exhibit a linear elastic compliance, i.e., a form of 'linear hardening' (see, e.g., Lehner & Kachanov, 1995). A shortcoming of (17) remains however the absence of an expected saturation level for work-hardening, that is a value of γ beyond which no further appreciable increase in cohesive strength will occur.

Using (17) in (8) and rearranging, the accumulated plastic shear strain may be expressed as a function of the loading parameters P and q as follows:

$$\gamma(\theta) = \frac{|q|}{c_2\cos\phi}\left\{\frac{1}{2}\sin(2\theta + \mathrm{Sign}(q)\phi) - \frac{c_1\cos\phi + (P - \frac{1}{2}q)\sin\phi}{|q|}\right\}\,. \tag{28}$$

This expression applies only if the right-hand side is positive, otherwise the internal variable $\gamma(\theta)$ is set to zero. Inserting it in (16) and carrying out the integration over the active angle range θ_1, θ_2, for which $\alpha^S(\theta) = 1$, one recovers the macroscopic permanent strain

$$[\epsilon^p_{ij}] = \frac{q}{2c_2\cos\phi}\begin{bmatrix} -1 & 0 & 0 \\ 0 & 2 & 0 \\ 0 & 0 & -1 \end{bmatrix} \times \Big| \mathrm{Sign}(q)\sin\phi\left(\tfrac{1}{3}\sin^3\theta - \tfrac{1}{5}\sin^5\theta\right)$$
$$- \cos\phi\left(\tfrac{1}{3}\cos^3\theta - \tfrac{1}{5}\cos^5\theta\right) - \frac{c_1\cos\phi + P\sin\phi}{3|q|}\sin^3\theta\Big|_{\theta_1}^{\theta_2}\,. \tag{29}$$

The terms within vertical bars are now evaluated for the limits of the active orientation range, θ_1 and θ_2, the former being subtracted from the latter. By use of (11) the plastic strain may then be expressed entirely in terms of the load parameters P and q.

To complete this discussion of triaxial deformation, let us now determine the contribution from a set of systematic joints or 'weak planes', initially present in the sample, to the overall plastic deformation. For simplicity, only a single joint set of orientation (θ_I, φ_I) is considered. If activated, its contribution to the macro-scale deformation will be the inelastic strain

$$[\epsilon^I{}_{ij}] = \tag{30}$$

$$\gamma_I(\theta_I) \begin{bmatrix} -\sin 2\theta_I \cos^2 \varphi_I & (1-2\cos^2\theta_I)\cos\varphi_I & -\sin 2\theta_I \sin 2\varphi_I/2 \\ (1-2\cos^2\theta_I)\cos\varphi_I & \sin 2\theta_I & (1-2\cos^2\theta_I)\sin\varphi_I \\ -\sin(2\theta_I)\sin(2\varphi_I)/2 & (1-2\cos^2\theta_I)\sin\varphi_I & -\sin 2\theta_I \sin^2\varphi_I \end{bmatrix},$$

where $\gamma_I(\theta_I)$ is obtained from expression (18) for $\gamma(\theta)$ upon replacing the argument θ by the fixed angle θ_I and using appropriate parameter values. The total permanent macro-scale deformation in the presence of the joint set is then given by the sum of (29) and (30).

In order gain some perspective, it will be of interest to evaluate the above general results for the following specific data set: A confining pressure P for the compressive test of 10 MPa, an initial cohesion c_1 on all potential discontinuities of 10 MPa, and a reduced initial cohesion one half the value of c_1 on a single preexiting joint set of orientation $(\theta_I, \varphi_I) = (\pi/4, 0)$. All discontinuity surfaces have the same coefficient of sliding friction $\mu = 0.6$, and the work-hardening constant c_2 is given the uniform value of 50 MPa.

Consider first the stress-strain response in triaxial compression, as given by (29) for a sample without weak joints of preferred orientation. First yield occurs in this test at a critical axial load, whose normalized magnitude $|\sigma_{22}|/c_1 = |-P+q_c|/c_1$ is found to equal 6.46 from (13) (cf. also Fig. 5). This marks the intercept with the zero-strain axis of the inelastic stress-strain response shown in Figure 6 (dashed line), in agreement with classical Mohr-Coulomb theory. The subsequent computed stress-strain response exhibits nonlinear hardening behaviour and this is clearly a consequence of the widening orientation range $[\theta_1, \theta_2]$ of active slip systems with increasing axial load, since individual systems have been assumed to respond linearly; the larger the range of activated discontinuities, the softer the material response.

The evolution of the angular range of activated slip systems with increasing plastic deformation is shown in Figure 7, in which the limiting angles θ_1 and θ_2 have been plotted versus the axial strain. The contribution of each activated plane to the macroscopic strain may be appreciated from the graphs shown in Figure 8 of the accumulated plastic shear on individual planes as a function of their orientation θ. Note the increase in the range of activated systems and in the magnitude of γ as loading proceeds.

The same triaxial loading history is now imposed on a specimen containing a set of weak planes with the above-assumed orientation and material properties. The computed macroscopic stress-strain response is given by the dotted line in Figure 6. Slip commences on the weak planes, but at the same normalized critical load of 6.46, found previously in the absence of weak planes, slip on distributed discontinuities sets in with a gradually expanding orientation range

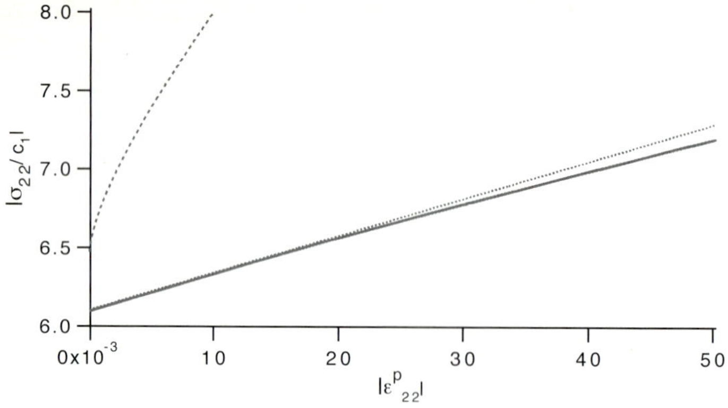

Fig. 6. Computed inelastic (axial) stress-strain response for triaxial compression in the absence (dashed line) and presence (solid line) of a single set of weak joints. Dotted line represents linear stress-strain response due to slip on weak joints alone.

of activated planes. Note that the slope of the stress-total strain curve remains constant as long as only the single set of weak joints is active (linear hardening), but decreases as soon as the growing range of distributed discontinuities becomes active and softens the response. When plotted versus total plastic strain, the orientation ranges of actively slipping discontinuities differ markedly in the absence (solid lines) and presence (dotted lines) of the weak joint set, the range being substantially diminished in the latter case (Fig. 7).

The second part of this section is concerned with a triaxial test in which the sample is allowed to undergo extension in the axial direction, so that both both sliding and opening displacements can be operative on potential surfaces of discontinuity. The representative volume element is assumed to be free of any weak joints. The macro-scale plastic deformation resulting from the opening mode is now discussed briefly and only the main results will be given.

The yield condition (12) gives the critical normal stress at which the opening mode of potential discontinuities is activated. On substituting the stress σ_N from (7) in this criterion, the resulting condition for the initiation of the opening mode is

$$-P + q\cos^2\theta - k_0(0) = 0\,. \tag{31}$$

It predicts that the first discontinuities will open at a critical value of q equal to $P + k_1$, at an orientation angle $\theta = 0$, i.e., in planes perpendicular to the x_2-axis (cf. Fig. 4b). Since work-hardening has been assumed for the opening mode as well, further loading results in the opening of discontinuities sub-parallel to the x_1, x_3-plane in a manner similar to the activation of the sliding cracks. The angle range of open discontinuities is given by the following solution to (31):

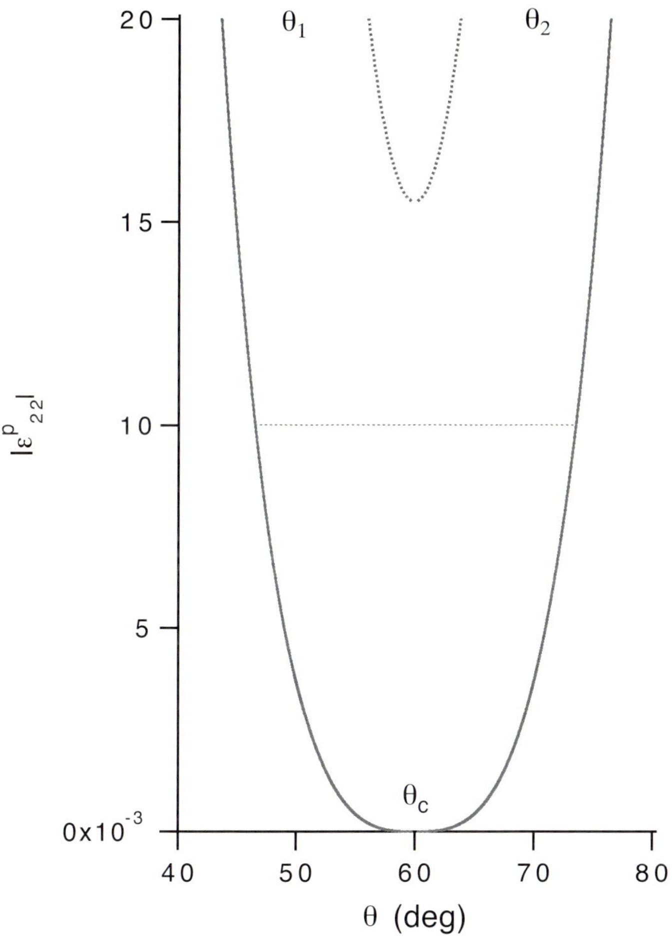

Fig. 7. The angular range of activated discontinuities in a compressive triaxial test as a function of the total plastic strain in the absence (solid line) and presence (dotted line) of a set of weak joints. Left and right branches showing θ_1 and θ_2, respectively.

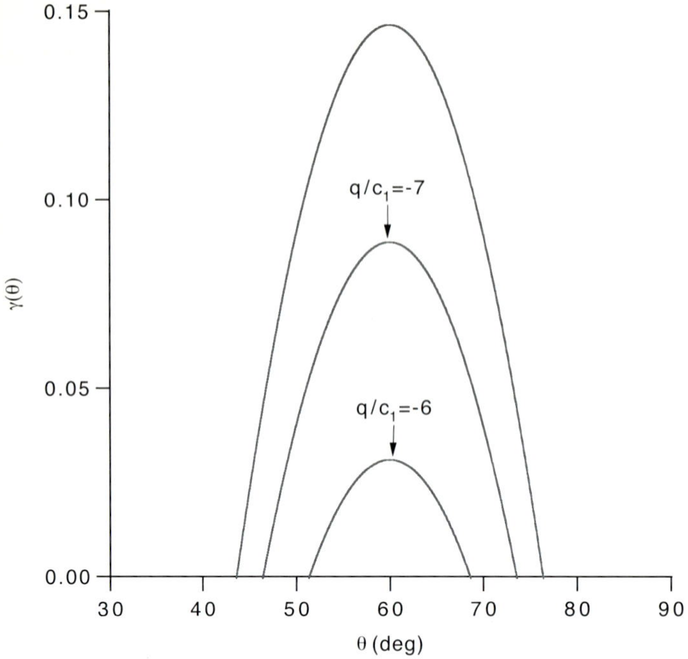

Fig. 8. Distribution of slip along activated discontinuities at different stages of the loading during the compressive triaxial test and in the absence of any plane of weakness.

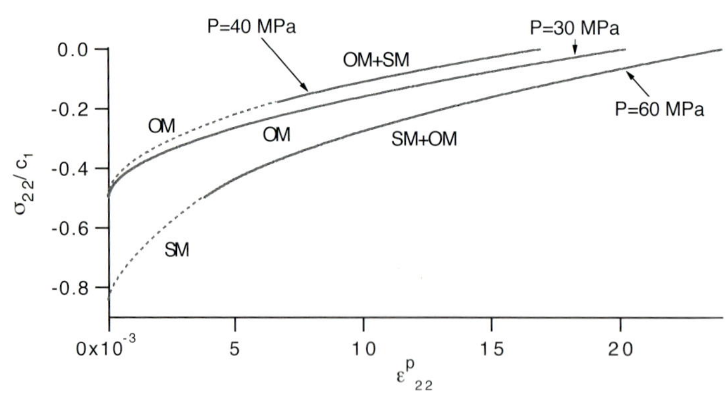

Fig. 9. Computed inelastic (axial) stress-strain response in triaxial extension for various values of the confining pressure P, showing that activation of shear- and opening modes, SM and OM, is sensitive to P and that both mechanisms can contribute simultaneously to the deformation.

$$\theta_3 = \arccos\left(\sqrt{\frac{k_0(0) + P}{q}}\right), \qquad q > 0 \,. \tag{32}$$

The macroscopic permanent strain associated with the opening of distributed discontinuities is found by integrating the appropriate term in expression (15) over the active orientation range and over time, giving

$$[\epsilon^p{}_{ij}] = \int_0^{\theta_3} \frac{d(\theta)}{2} \sin\theta \begin{bmatrix} \sin^2\theta & 0 & 0 \\ 0 & 2\cos^2\theta & 0 \\ 0 & 0 & \sin^2\theta \end{bmatrix} d\theta \,. \tag{33}$$

As with the sliding mechanism, work-hardening is characterized by a linear relation

$$k_0(d) = k_1 + k_2 d \,, \tag{34}$$

in which the two constants k_1 and k_2 are the same for all discontinuity orientations. This choice yields the expression $d(\theta) = q\cos^2(\theta) - P - k_1$ for the opening displacement. Substitution in (33) thus provides the following simple analytical solution for the strains

$$\begin{aligned} \epsilon^p_{11} = \epsilon^p_{33} &= \frac{1}{2k_2}\bigg| -\frac{1}{5} q(\sin^2\theta + \frac{2}{3})\cos^3\theta + \frac{1}{3}(P + k_1)\cos\theta(2 + \sin^2\theta)\bigg|_0^{\theta_3} \\ \epsilon^p_{22} &= \frac{1}{2k_2}\bigg| -\frac{2}{5} q\cos^5\theta + \frac{2}{3}(P + k_1)\cos^3\theta\bigg|_0^{\theta_3} \,. \end{aligned} \tag{35}$$

Since the two micro-scale deformation mechanisms operate independently, the total plastic deformation is obtained by adding up the individual contributions from slip (Eq. 29) and opening displacements (Eq. 35). Note again that the equations for slip have been written in a form that remains valid both for compression and extension.

The feature of interest in the extension test is the possibility of a simultaneous activation of the two micromechanisms of slip and opening and its consequence for the stress-strain response. This is illustrated by the examples shown in Figure 9. The parameters k_1 and k_2 of the hardening relation (34) are set to -5 and 10 MPa, respectively. Three extension tests are considered for confining pressures of 30, 40, and 60 MPa, corresponding to 3, 4, and 6 times the value of c_1. At the largest confining pressure P, yield occurs first by slip. From (12), the orientation of the discontinuity that slips first is found to be $\theta_c = \pi/4 - \phi/2$. Further loading increases the range of activated discontinuities until the normal stress becomes small enough to satisfy the yield criterion for the opening mode. Opening first occurs in planes normal to the x_2-axis when the axial stress reaches the value k_1, i.e., at $\sigma_{22}/c_1 = -0.5$ in the present example. Continued straining is accompanied by the simultaneous activation of opening and sliding modes. If the same test is conducted at the intermediate value for P of $4c_1$ (40 MPa), then the order of activation of the micromechanisms is reversed: The opening mode preceeds the sliding mode. According to (13), the latter is activated as soon as the axial stress attains the value $[2c_1 \cos\phi - (1 - \sin\phi)P]/(1 + \sin\phi)$, i.e., when $\sigma_{22}/c_1 = -0.1785$ for the above parameter values. Finally, for the smallest confining pressure P, equal to $3c_1$ (30 MPa), only the opening mode of deformation is activated. Note again the nonlinearity of stress-strain response, which results from the progressive opening of discontinuities sub-parallel to the x_1, x_3-plane.

This sequential activation of the two micro-scale deformation mechanisms and its sensitivity to the prevailing hydrostatic stress state is also an essential feature of plate bending. As will be discussed next, only slip will be activated in the plate intrados (region of compression), whereas slip and opening displacements can occur in the extrados (region of extension).

4 Bending of a plate

The initial stage in the formation of a fold is analyzed in this section as a problem of plate bending for a plate of thickness $2D$ and infinite extent (cf. Fig. 10a). The plate is initially subjected to an isotropic compressive stress and bending about the x_3-axis results in a linear variation across the plate thickness of the normal stress σ_{22}, as is illustrated in Figure 10b. The main difference between this situation and the previously discussed triaxial test lies in the spatial variation of the stress field. The two problems are in fact closely related and this should allow a quick grasp of the main results of this section.

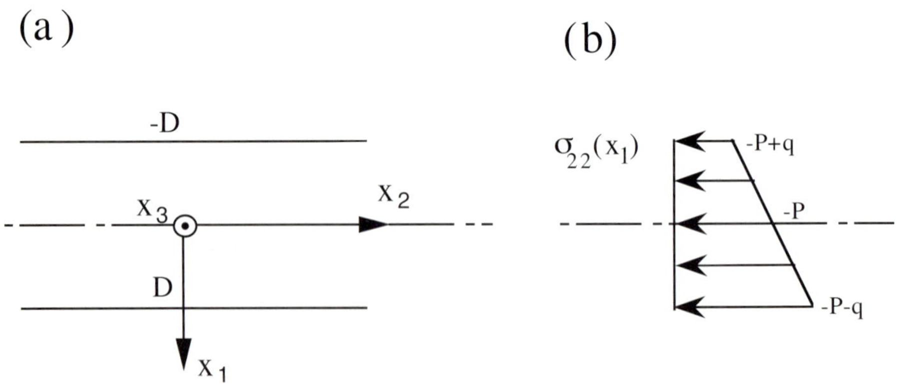

Fig. 10. (a) An infinite plate of thickness 2D, initially under a uniform pressure P. Bending about the x_3-axis generates principal stresses in the x_2 and x_3 directions that vary linearly across the plate thickness with a gradient proportional to the load parameter q, as illustrated in (b) for σ_{22}.

With reference to Figure , it is assumed that the deformation in every plane x_3=const. is one of plane strain. The unit normals $\boldsymbol{n}$ of all activated discontinuities must lie in the (x_1, x_2) plane, as a consequence of which the second Euler angle φ only takes on the values 0 and π. Moreover, the plane strain assumption and the absence of permanent deformation in the x_3-direction obviously implies that the elastic strains in that direction must also vanish and this constrains the component of stress σ_{33}. The state of stress in the plate is therefore given by:

$$\boldsymbol{\sigma} = -P\boldsymbol{I} - q\frac{x_1}{D}[\boldsymbol{e}_2 \otimes \boldsymbol{e}_2 + \nu \boldsymbol{e}_3 \otimes \boldsymbol{e}_3], \tag{36}$$

where P is the initial isotropic pressure and ν denotes Poisson's ratio. The stress is seen to vary linearly with x_1 across the plate and the parameter q is assumed to increase monotonically from $q = 0$ during loading. The stress distribution (36) satisfies equilibrium pointwise throughout the plate as long as gravity effects are disregarded. For $q \geq 0$, the bending bending generates compression in the region $x_1 > 0$ (the intrados region) and extension in the region $x_1 < 0$ (the extrados region). These stresses in the two regions thus differ in a similar way as in the triaxial compression and extension tests.

The components of the unit normal to potential discontinuities direction are again given by (6), with the second Euler angle φ set to either 0 or π. From (6) and (36), one proceeds as in the previous section to derive the following expressions for the normal stress and the resolved shear stress (for $q > 0$):

$$\sigma_N(\boldsymbol{n}) = -P - (qx_1/D)\cos^2\theta, \quad \tau(\boldsymbol{n}) = (q|x_1|/D)\sin\theta\cos\theta. \tag{37}$$

Substitution in the yield conditions (7) and (12) then produces the orientation ranges $[\theta_1, \theta_2]$ and $[0, \theta_3]$ of discontinuities activated in the shear and opening mode, respectively:

$$\begin{aligned} \theta_{1,2} &= \theta_c \mp \frac{1}{2}\left[\frac{\pi}{2} - \arcsin\left(\frac{2c_0(0)\cos\phi + (2P + qx_1/D)\sin\phi}{q|x_1|/D}\right)\right] \\ \theta_c &= \frac{\pi}{4} - \mathrm{Sign}(x_1)\frac{\phi}{2} \end{aligned} \tag{38}$$

and

$$\theta_3 = \arccos\left(\sqrt{\frac{k_0(0) + P}{q(-x_1)/D}}\right), \qquad (x_1 < 0). \tag{39}$$

As is seen from (38), slip occurs first when q reaches the critical value

$$q_c = \frac{2D(c_0(0) + P\sin\phi)}{|x_1|(1 - \mathrm{Sign}(x_1)\sin\phi)}, \tag{40}$$

and this clearly happens first in the extrados at $x_1 = -D$, while the initiation of slip at $x_1 = +D$ in the intrados requires a larger value of q (exactly by a factor 3 for a friction angle of 30 degrees).

It follows from (39) that the opening mode of potential discontinuities can be activated only in the extrados. This will happen first at $x_1 = -D$ for planes running parallel to the x_1, x_3-plane as soon as $q = k_0(0) + P$. Further loading results in the opening of sub-parallel discontinuities with orientation angles that lie within the range $[0, \theta_3]$. The value of the initial pressure P determines whether opening or slip will occur or whether both mechanisms are activated, as in the triaxial extension test. This effect is now illustrated by an example computation for three different values of P.

In this example, the plate is taken to be free of any weak joint or fracture systems of preferred orientation. All potential discontinuities have a friction coefficient μ of 0.3, a zero initial cohesion c_1 and a hardening coefficient c_2 of 50 MPa. The value $k_0(0)$ in (30), i.e. the constant k_1, is set to -10 MPa and the hardening parameter associated with opening, k_2, to 10 MPa. The plate half-thickness D is 500 m. Results are presented for a loading parameter q equal to 30 MPa and the three values of 30, 35 and 40 MPa for the confining pressure P.

In Figure 11a two plots are shown for the smallest confining pressure of 30 MPa, one for the extrados ($-1 \leq x_1 \leq -0.5$) and one for the intrados ($0.8 \leq x_1 \leq 1.$). The boundaries between the elastic core of the plate and the adjacent plastic zone are marked by a dashed line and only a portion of the core is shown. In the extrados, only the opening mechanism is found activated. Opening is initiated in a direction perpendicular to the x_1, x_3-plane at the limit of the elastic and plastic regions ($x_1 = -0.67D$) and the range of activated discontinuities increases away from the center of the plate. The limit θ_3 of the activated range is largest at the top surface of the plate. In the intrados, the range $[\theta_1, \theta_2]$ of slipping discontinuities exhibits a similar trend, beginning with a single critical orientation θ_c at elastic/plastic boundary and growing with increasing x_1.

The results obtained for the larger values of confining pressure of 35 and 40 MPa (Figs. 11b and 11c) remain qualitatively the same in the intrados, while differing markedly in the extrados. In the intrados, the main effect of a higher confining pressure is an enlargement of the elastic core of the plate. At 40 MPa the intrados remain entirely within the elastic range (Fig. 11c). In the extrados, on the other hand, increasing confining pressures are found to bring about a gradual shift from opening (dotted lines) to the shear (solid lines) as the preferred mode of plastic deformation, allowing both to spread simultaneously over different sections of the plate within a certain range of pressures (Fig. 11b). At the largest value of P, only the slip mechanism is found activated (Fig. 11c).

5 Concluding discussion

To obtain a first impression of the potential and the limitations of the simple model described in the foregoing, we now wish to to compare its predictions

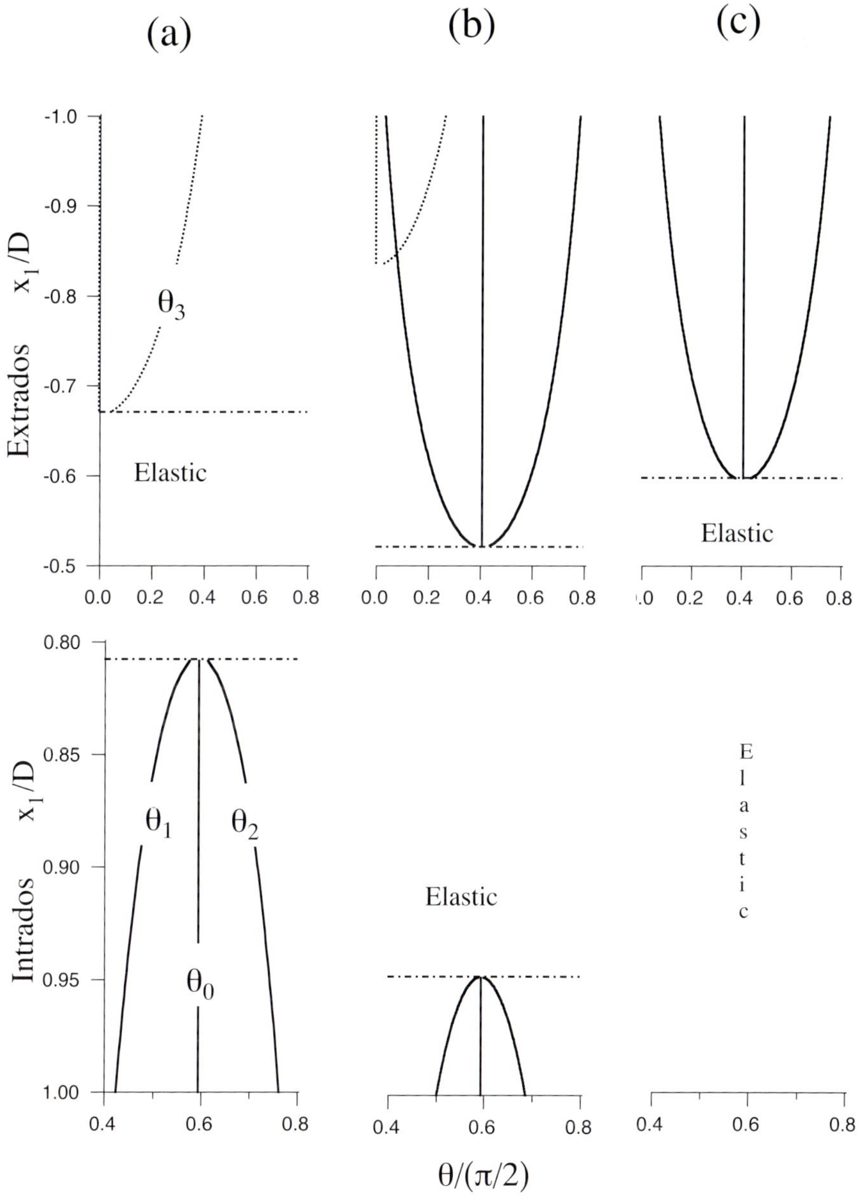

Fig. 11. Ranges of orientation of activated discontinuities in plate bending for different initial pressures P of (a) 30 MPa, (b) 35 MPa, and (c) 40 MPa. Upper plots for the plate extrados, lower plots for the intrados. Solid curves for the slip and dotted curves for the opening mode. Horizontal dashed lines mark the outer boundaries of the elastic core of the plate.

for the bending of a plate with the fracture pattern observed in a sandstone anticline outcropping in the Moroccan Anti-Atlas as shown in Figure 12a, which has kindly been provided by Gaulier *et al.* (1996).

The sandstone formation seen on Figure 12 is of Devonian age and is part of a fold formed during the Hercynian compression. The exposed sequence is at most 20 meters thick and is composed of several massive sandstone layers with thicknesses ranging from 2 to 5 meters. The study of Gaulier *et al.* (1996) provides a classification of the joint families in relation to the structural evolution of the folding. Fracture data, i.e., fracture orientation with respect to bedding, fracture density and spacing as well as fracture length and penetration depth, have been collected at more than 30 measurement sites. The classification of the joint families yields three fracture sets, two of them being apparent on Figure 12a. A first set, considered to predate the folding event, is made up of systematic joints perpendicular to bedding with a marked orientation. They can be detected on Figure 12a on the left-hand side of the photograph thanks to the sunlight's orientation: the exposed surfaces between two shadows belong to the systematic regional set. A second family is composed of extrados fractures perpendicular to the bedding and parallel to the fold hinge that is sub-perpendicular to the plane of the cliff. A precise description of this family is now given starting from the top of the anticline and moving down the exposed strata.

Looking at the top few meters of the anticline in cross-section, we note a dense population of meter-scale extrado joints. Their orientation, as plotted in stereographic projection in in Figure 12b (pole to fractures and fault plane trace) have been measured from the top horizontal surface along a distance of 10 meters for each diagram. The fracture dip and strike show a small scatter around the vertical and fold hinge direction, respectively. The density of meter-scale extrados joints is larger in some intermediate layers than in the top layer. This observation is often explained by a difference in mechanical properties of the various layers. Note that some of the joints are intersecting several sandstone strata. These discontinuities could result from the coalescence of fracture surfaces initiated in adjacent beds. They are often curved, dipping towards the center of the photograph. Furthermore, detailed field observations clearly indicate that some discontinuities have been activated successively as normal and reverse faults during folding. The reverse faulting event is disregarded from the subsequent discussion, since it belongs to the late stage of the compression. A typical normal fault, shown in Figure 13, is arrested as it crosses a layer with a different lithology. The secondary fracture pattern with a horse-tail geometry in the bottom layer permits to locate the fault boundary and to appreciate the extent of the extensional area.

We now turn our attention to a comparison between the predictions made by the constitutive model proposed in this paper and the field observations that have just been summarized. To begin with, it is necessary to fix the size of the elementary volume on the length scale of the field structure. We shall assume the elementary volume to cover the 20 metres of exposed thickness of the anticline. Motivated by the occurrence of joints that have been activated in

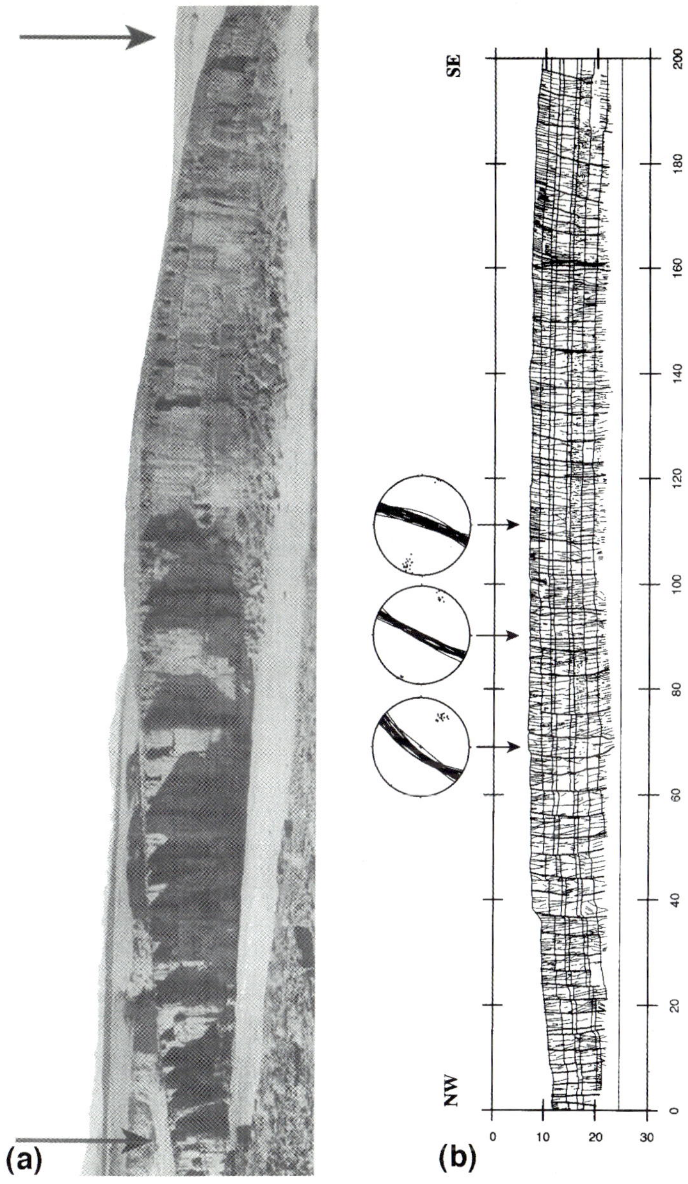

Fig. 12. The fracture pattern studied is from a Devonian sandstone anticline outcropping in the area of Tata-Akka, South Morocco. (a) General view of the exposed strata cross-cut by a Wadi in the direction NW-SE with a maximum height of 20 m. (b) Interpretation of the fracture pattern observed in (a) between the two arrows. Stereographic projections of fracture orientation in extrados (pole to fractures and fault plane trace) are also provided at three locations on the top of the anticline, separated by a distance of 10 metres. (Courtesy Gaulier *et al.*, 1996).

Fig. 13. Photograph of an extrado fracture that has been reactivated as a normal fault. It terminates in a layer composed of thin beds as a secondary fracture array with a 'horse-tail' geometry.

an opening mode, we interprete the elementary volume exposed in the field as representing a location along the vertical axis of Figure 11 in the extrados. The predicted orientations for discontinuities activated in the opening mode are then in qualitative agreement with those recorded on the stereonets of Figure 12b. The sliding mode is also found to occur simultaneously with the opening mode in the field, suggesting a situation as found for the extrados in Figure 11b at an intermediate pressure level. However, the Coulomb angle predicted by (12) for a friction coefficient of 0.3 is $\theta_c = \pi/4 + \phi/2 \approx 53^o$, which is very different from the subvertical orientation of the sheared discontinuities in the field. It would appear therefore, that in contrast with multiple faults of the classical Coulomb-type (cf. Mandl, 1988, p. 120 and Fig. I.3-23), 'normal faulting' in the present exposure has resulted from the initiation and coalescence of extrado joints. Their relatively narrow spacing, which appears to be controlled by bed thickness, has allowed the activation of a 'bookshelf mechanism' (Mandl, 1987). This is still recognizable in the left portion of Figure 12b, even though the faults have subsequently been reactivated in reverse and rotated clockwise. Rotation must have been anti-clockwise during 'normal faulting', however. The deformation seen in the field also comprises a ductile component in some of the lithologies present, an aspect that is clearly beyond the scope of our plasticity model in its present form.

In conclusion it may perhaps be said that both the strength and weakness of the proposed continuum model come from its simplicity. A major difficulty remains the justification of our use of simple yield criteria and hardening laws as a way of lumping together the description of complex processes of activation and growth of discontinuous or localized deformation on a finer scale. A second principal difficulty has to do with geometrical simplifications and the choice of scale. As illustrated by the example of the complex horse-tail geometry shown in Figure 13, there are always important features that can only be studied in isolation and on their own proper scale, features that fail to be representative even as an element of a larger ensemble of like elements. On the other hand, given suitable elements of this kind, such as the weak joint systems discussed in this paper, there remains the important task of addressing the scale effects that have been mentioned here only in passing, i.e., when does one have a *representative* elementary volume and how to deal with situations in which no statistically stationary distributions of the envisaged structural elements exist? In view of this and the above fundamental question about the lumped mechanical behaviour of discontinuities it seems clear that simple models of the present kind should possess a significant potential for further development.

References

Asaro, R.J. (1983). Micromechanics of Crystals and Polycrystals. *Advances in Applied Mechanics*, 23, 2–111.

Batdorf, S.B. and Budiansky, B. (1949). A mathematical theory of plasticity based on the concept of slip, *U.S. N.A.C.A. Technical Note 1871.*

Bott, M.H.P. (1959). The mechanics of oblique slip faulting. *Geol. Mag.*, 96, 109–117.

Bažant, Z.P. and Oh, B.H. (1985). Microplane model for progressive fracture of concrete and rock. *J. Eng. Mech., A.S.C.E.*, 111, 559–582.

Drucker, D.C. (1951). A more fundamental approach to plastic stress-strain relations. First U.S. Nat. Congr. Appl. Mech., 487–491.

Dunne, W.M. and Hancock, P.L. (1993). Paleostress analysis of small-scale brittle structures, In: *Continental deformation*, edited by P.L. Hancock, Chap. 5, pp. 101–120, Pergamon Press.

Gaulier, J.M., Mengus, J.M., Letouzey, J., Lecomte, J.C., Sassi W., Cacas, M.C. and Petit, J.P. (1996). Caractérisation des familles de diaclases formées au cours d'un phase de plissement. Cas des plis hercyniens de l'anti-Atlas marocain. Internal Report, Institut Français du Pétrole, Rueil-Malmaison.

Gauthier, B. and Angelier, J. (1985). Fault tectonics and deformation: a method of quantification using field data. *Earth and Planetary Science Letters*, 74, 137–148.

Hafner, W. (1951). Stress distributions and faulting. *Geol. Soc. Am. Bull.*, 62, 373–398.

Horii, H. and Nemat-Nasser, S. (1983). Overall moduli of solids with microcracks: Load induced anisotropy. *J. Mech. Phys. Solids*, 31, 155-1-71.

Jaeger, J.C. (1960). Shear failure of anisotropic rocks, *Geol. Mag.*. 97, 65–72.

Kachanov, M.L. (1982). A microcrack model of rock inelasticity. Part I: frictional sliding on microcracks. *Mechanics of Materials*, 1, 19–27.

Kachanov, M.L. (1982). A microcrack model of rock inelasticity. Part II: propagation of microcracks. *Mechanics of Materials*, 1, 29–41.

Koiter, W.T. (1953). Stress-strain relations, uniqueness and variational theorem for elastic-plastic materials with a singular yield surface, *Quarter. Appl. Math.*, 11, 350–354.

Lehner, F.K. and Kachanov, M. (1995). On the stress-strain relations for cracked elastic materials in compression. *Mechanics of Jointed Faulted Rock*, edited by H.-P. Rossmanith, Balkema, Rotterdam.

Leroy, Y.M. and Triantafyllidis, N. (1999). Stability analysis of incipient folding and faulting of an elasto-plastic layer on a viscous substratum. This Volume.

Mandel, J. (1964). Les conditions de stabilité et postulat de Drucker. In: *Rheology and Soil Mechanics*, IUTAM Symposium Grenoble, edited by J. Kravtchenko and P.M. Sirieys, pp. 58–68.

Mandl, G. (1987). Tectonic deformation by rotating parallel faults - the 'bookshelf' mechanism. *Tectonophysics*, 141, 277–316.

Mandl, G. (1988). *Mechanics of tectonic faulting.* Elsevier, Amsterdam.

Molnar, P. (1983). Average regional strain due to slip on numerous faults of different orientations. *J. of Geophys. Res.*, 88/B8, 6430–6432.

Nelson, R.A. (1985). *Geologic analysis of naturally fractured reservoirs.* Contribution in Petroleum Geology & Engineering, Gulf Publishing Company.

Nieuwland, D.A. and Walters, J.V. (1993). Geomechanics of the South Furious field. An integrated approach towards solving complex structural geological problems, including analogue and finite-element modelling. *Tectonophysics*, 226, 143–166.

Oertel, O. (1965). The mechanism of faulting in clay expriments. *Tectonophysics*, 2, 343–393.

Odé, H. (1960). Faulting as a velocity discontinuity in plastic deformation. In: *Rock Deformation*, edited by D. Griggs and J. Handin, Geol. Soc. Am. Memoir 79, pp 293–321.

Pan, J. and Rice, J.R. (1983). Rate sensitivity of plastic flow and implications for yield-surface vertices. *Int. J. Solids Structures*, 19, 973–987.

Petit, J.P., Auzias V., Rawnsley, K. and Rives, T. (1999). Development of joint sets in association with faults. This volume.

Reches, Z. (1983). Faulting of rocks in three-dimensional strain-fields II. Theoretical analysis. *Tectonophysics*, 95, 133–156.

Rawnsley, K.D., Rives, T., Petit, J.P., Hencher, S.R. and Lumsden, A.C. (1992). Joint development in perturbed stress fields near faults. *J. Struct. Geol.*, 14, 939–951.

Rudnicki, J.W. and Rice, J.R. (1975). Conditions for the localization of the deformation in pressure-sensitive dilatant materials. *J. Mech. Phys. Solids*, 23, 371–394.

Sanders, J.L. (1954). Plastic stress-strain relations based on linear loading functions. In: *Proc. Second US Nat. Congress of Applied Mechanics*, edited by P.M. Naghdi, ASME, 455-460.

Wallace, R.E. (1951). Geometry of shearing stress and relation to faulting. *J. Geol.*, 59, 118–130.

Zienkiewicz, O.C. and Pande, G.N. (1977). Time-dependent multilaminate model of rocks - A numerical study of deformation and failure of rock masses. *Int. J. Num. Anal. Methods in Geomechanics*, 1, 219–247.

Stability Analysis of Incipient Folding and Faulting of an Elasto-Plastic Layer on a Viscous Substratum

Yves M. Leroy[1] and Nicolas Triantafyllidis[2]

[1] Laboratoire de Mécanique des Solides Ecole Polytechnique, U.M.R. C.N.R.S. no. 7649 91128 Palaiseau Cedex, France
[2] Department of Aerospace Engineering, The University of Michigan, Ann Arbor, MI 48109-2140, U.S.A.

Abstract. The initiation of two modes of instability, viz. folding and faulting, is investigated theoretically for an elasto-plastic, frictional- cohesive layer which is underlain by a viscous substratum. The destabilizing factors that are allowed to come into play are the tectonic stress, buoyancy forces resulting from a gravitationally unstable density stratification, and the redistribution of material at the top surface by erosion and deposition processes. The bending stiffness of the overburden, a non-linear function of stress, has a stabilizing influence. A variational formulation of the stability problem allows one to detect the onset of global modes of instability, such as folds and surface modes for compression as well as neck-type modes for extension. Predictions of these global modes remain valid as long as the local condition of strong ellipticity is satisfied. Failure of this condition marks the onset of discontinuities in the velocity field, which are characteristic of localized faulting. The sensitivity of these predictions to the assumed behaviour of the overburden and substratum materials is explored for a prototype representative of a dip section through the Campos salt basin on the Brazilian continental margin. These results illustrate the importance of a proper selection of analogue materials for the design of physical laboratory models. This point is underscored by employing a deformation theory of plasticity which could be seen to reproduce in a simple manner the accommodation of bulk deformation by slip along a population of pervasive small faults in sedimentary rocks as well as in analogue materials, such as sands, used in the laboratory. A historical account of the use of deformation theories of plasticity in stability analyses and a derivation of the relevant incremental moduli are also given in this paper.

1 Introduction

Recent theoretical studies of the occurrence of two competing modes of instability, viz. folding in compression and localized faulting, have revealed the sensitivity of stability predictions for layered geological systems to the details of an assumed elasto-plastic, frictional-cohesive constitutive description (Leroy &

Triantafyllidis, 1996). In particular, the use of a deformation theory of plasticity, which mimics the presence of pervasive small faults, was found to yield predictions of realistic stress magnitudes for the onset of localized faulting as well as folding in compression (Triantafyllidis & Leroy, 1997). The present paper summarizes these results. It has been written with the further objective to motivate the use of a deformation theory of plasticity in stability analyses and to derive the relevant tangent moduli that relate stress and strain rates and are needed in such analyses. Finally, the paper focuses on the use of physical analogue experiments aimed at reproducing these types of instabilities in the laboratory. The sensitive dependence of such results on the choice of model material is emphasized in this context.

Plasticity theories have long been in use for modeling the onset of localized faulting in geological materials. Building upon a body of theory that had its roots in metal plasticity and the mechanics of granular materials, Odé (1960) proposed an interpretation of faults in sedimentary rocks in terms of a classical slip-line theory for rigid-perfectly plastic materials. In such a theory, the yield limit of the material corresponds to a surface of fixed position and shape in stress space. Although this slip-line theory has seen a substantial development for frictional-cohesive or Coulomb-plastic materials, especially for plane problems (Spencer 1964, 1971, 1982; Mandl & Fernandez Luque, 1970), probably the greatest benefit has been derived from it in situations in which the state of stress is fully determined by the equations of equilibrium and the yield condition, i.e., for the plastic limit equilibrium problems encountered typically in soil mechanics or with granular materials in general. The hyperbolic equations governing such limit states of stress were first obtained by Kötter in 1903 and there exists a substantial literature on methods of solution (see, e.g., Sokolovski, 1960). The stress characteristics obtained from these solutions are identified with slip lines, that is potential surfaces of discontinuity of a plane velocity field, or faults. The simplest, spatially uniform slip line patterns are associated with the so-called active and passive Rankine states in a gravitating halfspace and these are of course known to coincide with the classical Andersonian fault types. A comprehensive presentation of this approach to tectonic faulting is to be found in the monograph by Mandl (1988).

The prediction of incipient faulting based on rigid, perfectly-plastic models may be improved by allowing for the elastic and work hardening response of real materials. Here the inelastic part of the deformation is assumed to accumulate under increasing load in accord with some experimentally determined function. The yield surface, i.e., the boundary of the elastic domain thus changes its shape and/or position during deformation. The onset of localized faulting does not occur at first yield, however, as with rigid-perfectly plastic materials, but only in the non-linear range after the accumulation of a certain amount of inelastic deformation at increasing loads; it is marked by the loss of ellipticity of the governing partial differential equations (e.g., Rice, 1976). It is this type of plasticity model that forms the basis of our stability analysis.

While elasto-plasticity theories have met with much success in predicting the onset of localized faulting, they have only been brought recently to the study of diffuse folding or necking modes of instability in layered geological systems. For this class of problems the most frequently assumed constitutive model for rocks has been that of a viscous material (cf. the work of Biot, 1961 and Ramberg & Stephansson, 1964). This idealization permits folding instabilities to be captured but provides no insight into the occurrence of localized faulting, since it lacks the requisite destabilizing ingredient (Poirier, 1980). Triantafyllidis & Lehner (1993) were among the first to employ an elasto-plastic constitutive description in studying the stability of an initially horizontal interface between a solid and a fluid half-space, as determined by tectonic stress, buoyancy, and an initial stress gradient perpendicular to the interface. Triantafyllidis & Leroy (1994) extended their results by introducing a plasticity model appropriate for frictional and cohesive rocks and by allowing the overburden thickness to remain finite. In continuing this line of research, Leroy & Triantafyllidis (1996) provided a variational formulation of the stability problem for a substratum of finite thickness, which is valid for stable or unstable equilibrium states rather than for the neutral stability conditions alone. An asymptotic solution was subsequently developed by Leroy & Triantafyllidis (1998) that generalizes the classical solution for plate buckling (cf. Smoluchowski, 1909 and Ramberg & Stephansson, 1964) by allowing for the presence of a depth-gradient in the stress as well as for frictional-cohesive properties of the overburden. This solution is found to remain valid up to values of an appropriate 'small parameter' (defined by the product of the perturbation wavenumber and the overburden thickness) as large as 0.3 to 0.4 (Triantafyllidis & Leroy, 1997).

Whatever the mode of instability, localized faulting or diffuse folding (or necking), it is important to bear in mind that the solid-mechanics literature contains numerous studies indicating a sensitive dependence of stability predictions for these two modes on the details of the assumed constitutive behaviour. It is also known that the use of certain 'deformation theories' of plasticity tends to lower predicted critical loads for the onset of localized faulting (Rudnicki & Rice, 1975) and diffuse modes (Stören & Rice, 1975) of instability. These predictions are often found to be more realistic than those obtained with a classical 'flow theory' of plasticity that invokes a smooth yield surface (e.g., Hutchinson, 1974).

The success of deformation theories (with smooth yield surfaces) is due to a common feature with flow theories of plasticity based on a yield surface with a corner at the loading point. The existence of corners in the yield surface may be motivated by appeal to an underlying physical mechanism of plastic deformation involving the selective activation of favourably oriented micro-scale slip systems. In elasto-plastic polycrystalline materials, for example, the domain of elasticity in stress space is delimited by a yield surface that exhibits a corner or vertex at the load point, as a result of the simultaneous activation of slip along crystallographic planes of different orientation (Hill, 1967). It has been suggested that slip along populations of fissures or faults in a rock mass should give rise to the same feature

(Rudnicki & Rice, 1975). Detailed micromechanical analyses of the deformation behaviour resulting from the presence of frictionally sliding cracks in an elastic solid have indeed shown that the macroscopic yield surface develops a vertex at the load point in stress space (see, e.g., Lehner & Kachanov, 1995). This vertex at the load point facilitates the rotation of the principal stress directions and the related triggering of instabilities. The instabilities could be of a diffuse necking-type, followed by the initiation of shear bands, as in the plane-strain tensile test of metals (Tvergaard *et al.*, 1981), or could arise in the form of a surface-mode facilitating the subsequent development of shear bands, as in the pure bending of a thick beam (Triantafyllidis *et al.*, 1982).

Historically, the use of corner theories in stability analyses was preceeded by the application of deformation theories with a smooth yield surface. Under the assumption of proportional loading to the current stress state, such a theory may be obtained by integrating the incremental equations of plastic flow. The incremental moduli are then the same as for a non-linear elastic model when the load increment is in the direction of the current loading. These moduli differ from the flow theory moduli for increments orthogonal to this direction through the introduction of a secant modulus into the deformation theory. Since the stiffness in the direction orthogonal to the current loading direction is reduced in the deformation theory, it becomes more conducive to the triggering of instabilities than the corresponding flow-theory of plasticity.

It should be noted that the validity of a deformation theory with smooth yield surface to test the importance of a yield vertex in linearized stability or bifurcation analyses is only justified if the loading remains close to proportional (Budiansky, 1959). With that hypothesis, the deformation theory has similar effects as the flow theory with a yield vertex, but it should be stressed that the two will, in general, not be equivalent. The equivalence of the incremental moduli of these two theories for arbitrary stress increments has only been assessed for the case of a polycrystalline aggregate from the results of calculations based on the self-consistent method (Hutchinson, 1970). The pertinence of deformation theories for fissured rock masses if the activation of all micro slip systems is continuously maintained during loading has been established by Lehner & Kachanov (1995) and remains to be generalized to 3D stress states.

The present paper is organized as follows. In the next section we introduce a model problem which we have analyzed previously in a series of papers (Triantafyllidis & Leroy, 1994; Leroy & Triantafyllidis, 1996; Triantafyllidis & Leroy, 1997). We shall emphasize the role of the in-situ stress distribution in the overburden and its link to the incremental stress-strain relation which enters the variational formulation of the stability problem and determines the non-linear stiffness of the overburden. The third section introduces the notions of flow and deformation theories of plasticity and shows the analogy between the deformation theory based on a smooth yield surface and a flow theory for a plasticity model defined for a yield surface having a corner. A physical motivation is given for the presence of a yield vertex in a plasticity model of rocks deforming on a tectonic scale. The complete derivation of the incremental moduli for the deformation theory necessary for a stability analysis is relegated to an Appendix. The fourth section contains an application of this type of stability analysis to a prototype representing a folded section through the Campos salt basin, offshore Brazil, studied by Cobbold & Szatmari (1991) and Demercian *et al.* (1993). To explain the difference in stability modes found in the field and observed in laboratory model experiments, we turn our attention to the requisite scaling rules and, in particular, to the proper choice of analogue materials. It is demonstrated that differences between the elasto-plastic deformation response of the postulated field prototype and that of the laboratory analogue material can account for the observed discrepancy in the modes of instability and hence, that an adequate understanding and scaling of material behaviour will be essential to the reproduction of the correct instability modes in the laboratory. Motivated by this observation, the last section of the paper concludes with a discussion of the need to develop constitutive relations for pervasively fractured rock masses that may be assumed to deform as a continuum on a tectonic scale, the subject of a companion paper in this volume (Leroy and Sassi, 1999).

2 The model problem

In this section we wish to introduce a model problem that was analyzed by the authors in the series of papers mentioned earlier, and to summarize the solution techniques developed there. The model geometry, the in-situ stress parameterization and the formulation of the stability problem are first discussed. Three solutions to the stability problem are then considered, viz. an approximate solution based on the finite-element method, an analytical solution valid for negligible gravitational effects, and an asymptotic solution for large wavelengths in comparison with an overburden thickness. These solutions all pertain to the appearance of diffuse modes of instability and are valid if a local condition of strong-ellipticity holds through-out the overburden. Failure of this condition signals discontinuities in the velocity field which mark the onset of localized faulting.

The geometry of the model problem considered is shown in Figure 1. The overburden and the buoyant substratum are layers of infinite horizontal extent with finite thicknesses denoted by H_a and H_b, respectively. In what follows, the suffixes a and b are used to denote field variables or parameters associated with the layer 'above' (overburden) and 'below' (substratum), respectively. The two layers are perfectly bonded along an initially horizontal interface. The substratum rests on a rigid basement with which it forms an interface that acts either as a perfect bond or else admits frictionless interlayer slip to occur. A coordinate system is introduced with the x_1- and x_3-axes in the plane of the overburden/substratum interface. The x_2 coordinate can be normalized with respect to the overburden and the substratum thickness, resulting in the introduction of the two dimensionless coordinates ξ and ζ, respectively. The unperturbed or initial state of stress in the overburden varies with depth and is characterized by

$$\begin{aligned} \overset{a}{\sigma}_{11}(\xi) &= \sigma_0 \cos(\varphi) + k_1 \overset{a}{\rho}\, g H_a \xi\,, \\ \overset{a}{\sigma}_{22}(\xi) &= \overset{a}{\rho}\, g H_a(\xi - 1)\,, \\ \overset{a}{\sigma}_{33}(\xi) &= \sigma_0 \sin(\varphi) + k_3 \overset{a}{\rho}\, g H_a \xi\,, \end{aligned} \tag{1}$$

in terms the four parameters σ_0, φ, k_1 and k_3. The vertical stress $\overset{a}{\sigma}_{22}$ results

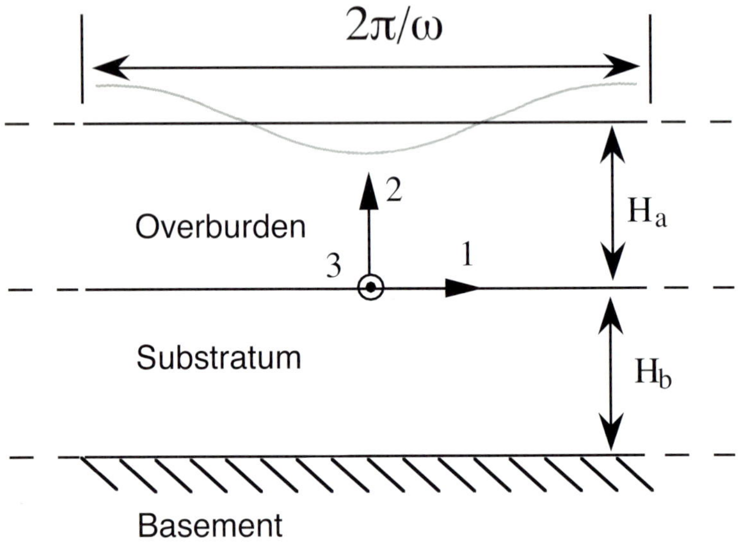

Fig. 1. The model problem comprises an elasto-plastic, frictional-cohesive overburden overlying a viscous substratum and a rigid basement. The contact between substratum and basement is either a perfect bond or will permit frictionless slip. Stability is analyzed for wavy perturbation of wavelength $2\pi/\omega$.

solely from the lithostatic pressure, $\overset{a}{\rho}$ denoting the local bulk density of the overburden and g the acceleration of gravity. The stresses in the x_1- and x_3-directions at the interface between the overburden and substratum ($\xi = 0$) are defined by the orientation angle φ and the scalar σ_0. Their depth gradients are determined by the coefficients k_1 and k_3 and the lithostatic pressure. These coefficients may assume any value between 0.8, commonly seen in sedimentary basins (Breckels & van Eekelen, 1982), and 2, as found in fold belts (McGarr & Gay, 1978). The substratum displays viscous, fluid-like behaviour, its density is lower than the density of the overburden material, and its unperturbed or initial in-situ stress distribution is hydrostatic and determined by its density and the weight of the overburden:

$$\overset{b}{\sigma}_{11}(\zeta) = \overset{b}{\sigma}_{22}(\zeta) = \overset{b}{\sigma}_{33}(\zeta) = -\overset{a}{\rho}\, gH_a + \overset{b}{\rho}\, g\zeta H_b\,. \tag{2}$$

The stability of the stratified system depends on the material stiffness of the cohesive overburden. This stiffness is defined locally in the incremental constitutive relation

$$\dot{\Pi}_{ij} = \overset{a}{L}_{ijkl}\dot{u}_{k,l}\,, \tag{3}$$

by the fourth-order tensor $\overset{a}{L}_{ijkl}$ which relates the first Piola-Kirchhoff stress rate $\dot{\Pi}_{ij}$ to the velocity gradient $\dot{u}_{k,l}$. In (3), the comma stands for partial differentiation with respect to the coordinate x_l. Repeated indices imply summation from 1 to 3. The definitions of the various stress tensors used so far can be found, for example, in Ogden (1984).

The relation between the moduli introduced in (3) and the stress parameterization (1) is now discussed. It is assumed that the overburden material is homogeneous and that its mechanical behaviour may be characterized by suitable laboratory experiments, yielding a plot of equivalent stress versus equivalent permanent strain such as presented in Figure 2. The equivalent stress is of a Drucker & Prager type (Drucker & Prager,1952): $\tau + \mu p$ a linear combination of the equivalent shear stress τ, viz. the square root of the second invariant of stress deviator, and the first invariant of stress tensor p times a scalar factor μ which is the friction coefficient. The equivalent permanent strain γ is a measure of the plastic deformation accumulated during the whole loading process. Its exact definition is provided further below. It is now assumed that the state of stress at every point throughout the overburden defined in (1) is reached by proportional loading: the ratio of principal stresses and their orientation is constant during the loading process. In that instance, one can compute from (2) the equivalent stress $\tau + \mu p$ and deduce the equivalent strain γ from the laboratory test results sketched in Figure 2. The value of this equivalent strain is unique in view of the monotonically increasing relation shown there. Its knowledge permits one to construct the multi-axial incremental response (3) at any point in the overburden for an appropriate plasticity model (cf. the next section and the Appendix on the Rudnicki-Rice model).

Three of the four factors that will affect the stability of our two-layer model have been considered so far. Among these, the (positive) density difference between the overburden and substratum and the tectonic compressive stress σ_0 have a destabilizing influence, while the third factor, i.e., the stiffness of the overburden material tends to stabilize the system. The last factor to be accounted for in this analysis is the redistribution of material at the surface which will be allowed for, following Biot & Odé (1965), by stipulating the removal by erosion of any material that is uplifted with respect to a spatially fixed surface, and the deposition of new sediment wherever the material tends to subside below that level. This redistribution condition will act as a destabilizing factor. It is

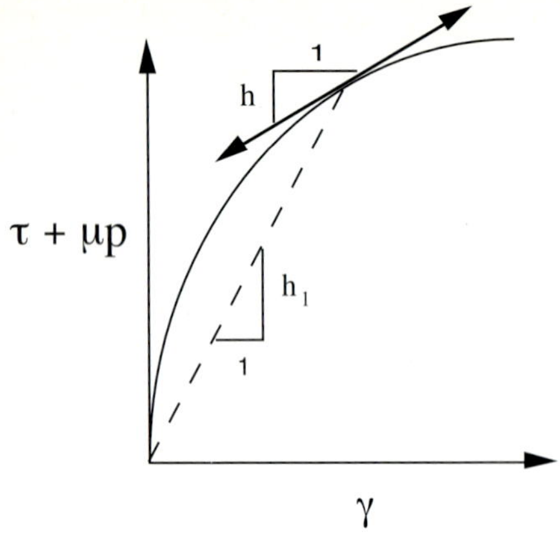

Fig. 2. Schematic representation of an experimental equivalent stress - permanent strain relation that permits one to define the scalar tangent and secant moduli h and h_1, respectively

expressed through the following mixed-boundary condition in the traction and displacement rate:

$$\dot{\Pi}_{2j} = \overset{a}{\rho}\, g \dot{u}_2 \delta_{j2} \,, \tag{4}$$

in which δ_{ij} is the Kronecker delta symbol ($\delta_{ij} = 1$, if $i = j$ and $\delta_{ij} = 0$, if $i \neq j$).

The stability of this layered system, which sustains a three-dimensional initial state of stress has been analyzed by the authors for two-dimensional perturbations with wavelength $2\pi/\omega$ in which ω is the wavenumber, Figure 1. The superposed perturbation are assumed to satisfy the simplifying assumption of plane-strain. The variational formulation of the stability problem, which is obtained by first integrating the linearized rate-form of the equilibrium equations over the substratum and the overburden, accounting for the boundary conditions at the layer interfaces and top surface as well as the constitutive relations, is presented in Leroy & Triantafyllidis (1996). The reader is referred to their work for the details of the derivation and for the exact expression of the variational equality (Eq. 17). It constitutes a generalized eigenvalue problem for the stability exponent Λ, having the dimension of stress, which determines the rate of growth or decay with time of the perturbation. A positive real part of Λ signals an instability for a mode with wavenumber ω which is the eigenvector of the variational equality. Neutral stability is marked by a zero real part of Λ. A negative real part for all admissible eigenmodes is necessary to have linear stability.

The variational equality is only applicable to sufficiently smooth, diffuse modes. Hence, one must verify that the stress state in the overburden is such that initiation of any local type of instability in the form of a shear band, i.e.

localized faulting, is precluded. Consequently, the stability calculations comprise two steps: the first consists of determining if the stress state satisfies the strong ellipticity condition for all points in the overburden. Equation 79 of Triantafyllidis and Leroy (1994) is used for that purpose. If this condition is not met, a local mode of instability in the form of localized fault has been detected. If this condition is met, the second step of the stability analysis consists of a study of smooth modes with the variational equality discussed above.

No analytical solutions to the variational equality were obtained because of the stress gradient present in the overburden and the non-linear mechanical response. It is only for the particular case of zero gravity that such an analytical solution could be derived (Triantafyllidis and Leroy, 1997). This solution is of interest if the dimensionless rate of growth of the instability Λ/G, in which G is the modulus of elasticity in shear, is not negligible compared to one, as illustrated for the case of an elastic beam on a viscous foundation by Leroy and Triantafyllidis (1996). Another procedure to construct a solution to the variational problem is to propose an asymptotic development for long wavelength compared to the overburden thickness (Leroy and Triantafyllidis, 1998). The small parameter is then ωH_a. Such an asymptotic solution generalizes the classical elastic plate solution on a viscous foundation considered by Smoluchowski (1909), Biot (1961) and Ramberg and Stephansson (1964) in several ways: the overburden has a stiffness which is a function of stress and stress-gradient and the finite-thickness of the substratum are accounted for. That asymptotic solution is found to be accurate even for large values of the small parameter such as 0.4. That range of validity was estimated with a numerical solution to the variational problem, based on the finite-element method, which approximates the unknown exact solution with the appropriate accuracy.

3 Constitutive relations

The stability predictions to be presented in the next section rely on the provision of moduli introduced in equation (3) which relate strain and stress rates. For example, these moduli are known for the flow- and deformation-theory versions of the Rudnicki-Rice plasticity model, which is based on a smooth yield surface in stress space, and are used for all the calculations to be reported. This section provides the definition of the two theories and show how a deformation-theory version can be constructed from any flow-theory version of a plasticity model. Furthermore, a comparison is made between the deformation theory and a corner theory to motivate the use of the former in a stability analysis. This section has the same objectives as in the work of Rudnicki (1984), who included the pore-fluids effects disregarded in our analysis. Most of the arguments to be presented have been put forward earlier by Koiter (1953), Sanders (1954), Budiansky (1959), and Stören & Rice (1975). It is hoped that the extension of some of their results to frictional, dilatant rocks will not harm the simplicity and elegance of the original ideas.

3.1 Definition of flow and deformation theories

According to Hill (1950), the distinction between *flow* and *deformation* theories of plasticity made in the solid-mechanics literature is due to Ilyushin. Hill, however, recommends *incremental-strain* and *total-strain theories* as alternative names. The two types of theory are are also referred to by the names of the originators of their earliest versions as *Prandtl Reuss*- and *Hencky*-type plasticity theories. Their difference will now explained.

In adopting a small strain approximation, we consider the decomposition of the total strain $\boldsymbol{\epsilon}$ into a reversible elastic part $\boldsymbol{\epsilon}^e$ and a permanent plastic part $\boldsymbol{\epsilon}^p$, the former being related to the stress $\boldsymbol{\sigma}$ by Hooke's law (Bold-face quantities denote vectors and second-order tensors). A Hencky-type theory of plasticity relates the *total* plastic strain $\boldsymbol{\epsilon}^p$ to the current stress by

$$\boldsymbol{\epsilon}^p = \gamma_D \frac{\partial \psi}{\partial \boldsymbol{\sigma}} , \tag{5}$$

where ψ is a potential function of the stress invariants. The scalar γ_D depends also on the state of stress, but not on the loading path in stress space as long as a condition of continued loading remains satisfied.

Relation (5), together with Hooke's law, yields the total *strain* produced by the current stress. In the absence of any local episode of unloading or neutral loading, the relation could equivalently characterize the response of a non-linear elastic material.

For a Prandtl Reuss-type theory, it is the plastic strain *increment* or *rate* that is obtained from potential ψ, i.e.,

$$\dot{\boldsymbol{\epsilon}}^p = \dot{\gamma}_F \frac{\partial \psi}{\partial \boldsymbol{\sigma}} , \tag{6}$$

and the parameter $\dot{\gamma}_F$ now depends upon the stress and stress rate. Equation (6) governs the plastic *flow* of the material produced by an increment of stress that satisfies a condition of continuous loading.

The two theories coincide, if the ratio of the principal stresses and their orientations remain fixed during loading, a condition referred to as proportional loading. It follows that a corresponding deformation theory may be constructed from every flow theory and an assumption of proportional loading as it is shown next. Deformation theories are of limited interest for solving boundary value problems in view of the exclusion of unloading. However, as will be discussed here, they can be of great help for testing the sensitivity of stability predictions to the details of the microscale deformation mechanism. This point will become clear in the last part of this section.

3.2 A flow theory for frictional and cohesive materials

The second objective of this section is to illustrate how to construct from a flow-theory of plasticity with smooth yield surface a corresponding deformation-theory version. For that purpose, a flow-theory version of a plasticity model applicable to cohesive and frictional materials is first presented in details. For the sake of simplicity, the elastic deformation of the material is disregarded compared to the permanent deformation.

Consider a smooth yield function ϕ of the type introduced by Drucker & Prager (1952) for pressure-sensitive materials[1]:

$$\phi \equiv \tau + \mu p - g(\gamma) \leq 0\,,$$
$$\tau = \sqrt{\frac{1}{2}\boldsymbol{\sigma}'\!:\!\boldsymbol{\sigma}'}\,, \qquad \sigma = \frac{\mathbf{I}\!:\!\boldsymbol{\sigma}}{3}\,, \qquad \boldsymbol{\sigma}' = \boldsymbol{\sigma} - \sigma\mathbf{I}\,. \tag{7}$$

The yield function ϕ involves a dependence on the equivalent shear stress τ, the mean stress p (positive in tension), both defined earlier, and on the accumulated equivalent plastic strain γ through the work hardening function $g(\gamma)$. The equivalent strain is defined by

$$\gamma = \int_0^{\boldsymbol{\epsilon}} \sqrt{2\dot{\boldsymbol{\epsilon}}'\!:\!\dot{\boldsymbol{\epsilon}}'}\,, \tag{8}$$

and the work hardening function $g(\gamma)$ is assumed to be known from the results of an experiment as sketched in Figure 2. A power law is chosen here to parameterize this function.

Plastic flow does not take place and the material is infinitely stiff (since we have disregarded its elasticity), if ϕ is strictly negative. For plastic flow to occur, ϕ must remain strictly equal to zero. Plastic flow also requires the product between the stress rate $\dot{\boldsymbol{\sigma}}$ and the normal to ϕ in stress space to be positive

$$\mathbf{n}\!:\!\dot{\boldsymbol{\sigma}} > 0\,, \qquad \mathbf{n} \equiv \frac{\partial\phi}{\partial\boldsymbol{\sigma}} = \frac{\boldsymbol{\sigma}'}{2\tau} + \frac{\mu}{3}\mathbf{I}\,. \tag{9}$$

The orthogonality condition $\mathbf{n}\!:\!\dot{\boldsymbol{\sigma}} = 0$ defines a state of neutral loading. Since ϕ remains zero during plastic straining or neutral loading, its rate also vanishes, i.e, $\dot{\phi} = 0$. This so-called consistency condition allows the hardening rate $\dot{\gamma}$ to be expressed in terms of the stress rate by

$$\dot{\gamma} = \frac{1}{h}\mathbf{n}\!:\!\dot{\boldsymbol{\sigma}}\,, \tag{10}$$

[1] The prime denotes the deviator of the stress tensor and the same notation will be used for the deviators of other tensors. The components of the second-order identity tensor $\mathbf{I}$ are given by the Kronecker delta δ_{ij}. A colon between any two second-order tensors denotes the scalar product $\mathbf{A}\!:\!\mathbf{B} = A_{ij}B_{ij}$ in terms of the Cartesian components of the two tensors, summation over the range of a repeated index being understood as usual.

where h is the derivative of the function $g(\gamma)$, as it is illustrated in Figure 2.

The strain rate is derived from a potential function ψ in accord with the flow rule (6), the potential being given by

$$\psi = \tau + \beta\sigma\,, \tag{11}$$

where β is a dilatancy coefficient. Note that the coefficients of friction and dilatancy are taken constant for simplicity, a choice that cannot be rigorously justified. The direction of plastic flow is determined by the normal to the flow potential

$$\mathbf{m} \equiv \frac{\partial\psi}{\partial\boldsymbol{\sigma}} = \frac{\boldsymbol{\sigma}'}{2\tau} + \frac{\beta}{3}\mathbf{I}\,. \tag{12}$$

Note that the product $\mathbf{m}{:}\mathbf{m}$ turns out to be constant and equal to $1/2 + \beta^2/3$ so that the rate $\dot{\mathbf{m}}{:}\mathbf{m}$ must be zero. The significance of this feature will become clear in the course of the construction of our deformation theory. The formulation of the flow-theory plasticity model is now completed by a relation for strain rate in terms of the stress rate, which is obtained from the flow rule (6) and the consistency condition (10) in the following form[2]:

$$\dot{\boldsymbol{\epsilon}} = \frac{1}{h}\mathbf{m}\otimes\mathbf{n}{:}\dot{\boldsymbol{\sigma}}\,. \tag{13}$$

The simple flow theory that has just been summarized may now be given a geometric interpretation (cf. Figure 3a). The solid and dashed curves shown there represent the yield function ϕ and flow potential ψ at the current loading point, respectively. The projection of the stress rate $\dot{\boldsymbol{\sigma}}$ onto the normal to the yield surface $\mathbf{n}$ is positive, ensuring the occurrence of plastic flow (conditions in equation 9) and yielding a value for $\dot{\gamma}h$. This scalar determines the magnitude $\dot{\gamma}$ of the plastic strain $\dot{\boldsymbol{\epsilon}}^p$, but not its direction, the latter being constrained by (12) to coincide with the normal $\mathbf{m}$ to the flow potential. This important constraint is relaxed in a deformation theory, as is to be shown next.

3.3 Construction of a deformation theory

The deformation theory corresponding to the flow theory presented above is now constructed. The difference in the incremental moduli of the two theories will then be examined and will show clearly that, under given in-situ stress

[2] The dyadic or tensor product $\mathbf{a}\otimes\mathbf{b}$ of two vectors $\mathbf{a}$ and $\mathbf{b}$ constitutes a second-order tensor with components $(\mathbf{a}\otimes\mathbf{b})_{ij} = a_i b_j$; it is defined through its action on an arbitrary vector $\mathbf{c}$ by $(\mathbf{a}\otimes\mathbf{b})\,.\mathbf{c}=(\mathbf{c}.\mathbf{b})\mathbf{a}$, the single dot denoting the scalar or 'dot product' of two vectors. Later, we shall also require the product of two second-order tensors $\mathbf{C}=\mathbf{A}.\mathbf{B}$ with components $C_{ij} = A_{ik}B_{kj}$ in a Cartesian coordinate system.

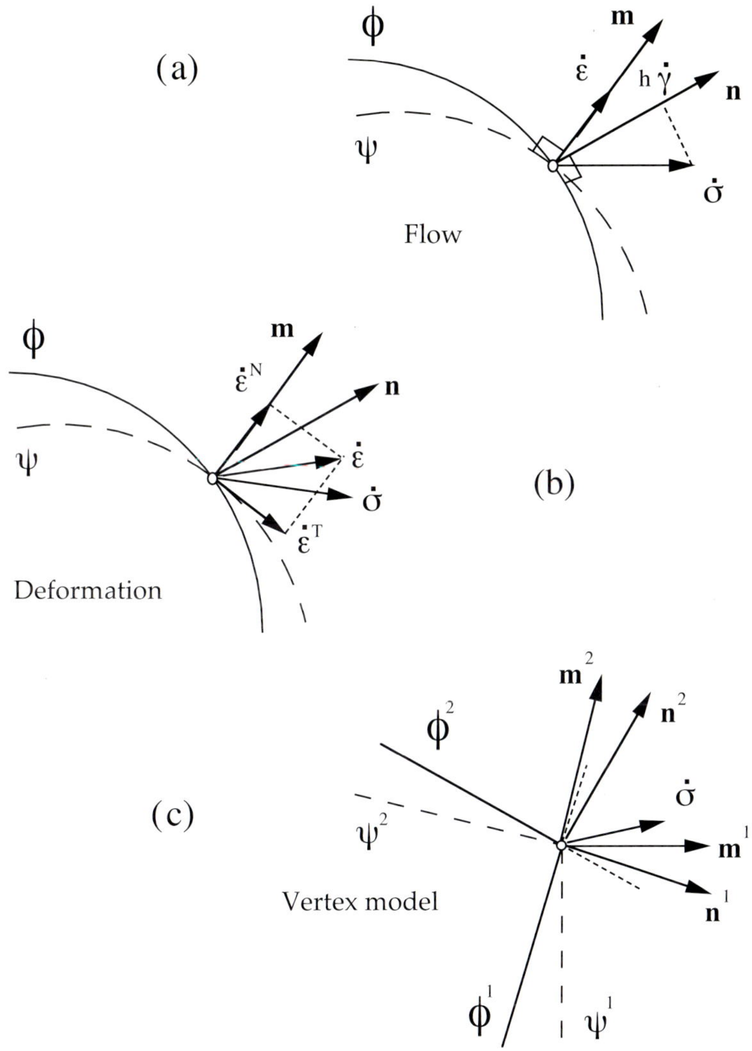

Fig. 3. Geometrical interpretation of the flow rule in stress space for the flow theory (a), for the deformation theory (b) and for a flow theory with a vertex in the yield surface (c). The solid and dashed curves represents the yield surface and the flow potential at the current stress point, respectively

conditions, a deformation theory tends to be more conducive to the appearance of an instability than its flow-theory counterpart.

To construct the deformation-theory counterpart to the flow theory presented, the current state of stress is assumed to have been reached by proportional loading. Such a stress path implies constant surface normals for the flow potential and yield function and we shall use the superscript $'o'$ to denote these fixed directions. If combined with (10), relation (13) assumes the simple form $\dot{\boldsymbol{\epsilon}} = \dot{\gamma}\mathbf{m}^o$ and from this the current strain is found in terms of the current stress by quadrature

$$\boldsymbol{\epsilon} = \gamma \mathbf{m}^o \,, \tag{14}$$

which is clearly a stress-strain relation of the type (5), characterizing a Hencky-type deformation theory.

The moduli of this theory, which enter the incremental relation (3) required for the stability analysis, have to be valid for any stress increment orientation and thus, the hypothesis of proportional loading for the subsequent increment in stress has now to be relaxed. Then, by differentiating expression (14), one obtains

$$\dot{\boldsymbol{\epsilon}} = \frac{1}{h}\mathbf{m} \otimes \mathbf{n}{:}\dot{\boldsymbol{\sigma}} + \gamma \dot{\mathbf{m}} \,, \tag{15}$$

where $\dot{\mathbf{m}}$ in the second term of the right-hand side is orthogonal to $\mathbf{m}$ and expressed by

$$\begin{aligned} \dot{\mathbf{m}} &\equiv \frac{\partial^2 \psi}{\partial \boldsymbol{\sigma} \partial \boldsymbol{\sigma}}{:}\dot{\boldsymbol{\sigma}} = \frac{1}{\tau}\mathbf{M}{:}\dot{\boldsymbol{\sigma}} \,, \\ \mathbf{M} &= (\mathbb{I} - \frac{1}{3}\mathbf{I} \otimes \mathbf{I})\frac{1}{2} - \frac{\boldsymbol{\sigma}'}{2\tau} \otimes \frac{\boldsymbol{\sigma}'}{2\tau} \,. \end{aligned} \tag{16}$$

The fourth-order identity tensor $\mathbb{I}$ has the components $\mathbb{I}_{ijkl} = 1/2(\delta_{ik}\delta_{jl} + \delta_{il}\delta_{jk})$. Combining (16) and (15), we obtain the incremental relation

$$\dot{\boldsymbol{\epsilon}} = [\frac{1}{h}\mathbf{m} \otimes \mathbf{n} + \frac{1}{h_1}\mathbf{M}]{:}\dot{\boldsymbol{\sigma}} \,, \tag{17}$$

where $h_1 = \tau/\gamma$ is a secant modulus, the significance of is also made clear by Figure 2.

Let us now consider the response of a specimen as predicted by the two types of theory for a proportional loading test up to the current stress of interest. This particular loading path may be thought of as providing a fundamental solution, the stability of which with respect to small perturbations is to be examined in a linearized stability analysis. The state of stress at a given instant during loading may be described by a direction tensor $\boldsymbol{\sigma}^o$ and an amplitude α, the latter being a scalar quantity that increases monotically during loading: $\boldsymbol{\sigma} = \alpha\boldsymbol{\sigma}^o$. Decomposing $\boldsymbol{\sigma}^o$ into a spherical and a deviatory part, one finds that the stress rate $\dot{\alpha}\boldsymbol{\sigma}^o$ is orthogonal to the tensor $\mathbf{M}$, viz. $\mathbf{M}{:}\boldsymbol{\sigma}^o = 0$. From (13) and (15) it is apparent that the strain rate in the direction of the normal $\mathbf{m}$, which is denoted by $\dot{\boldsymbol{\epsilon}}^N$, Figure 3b, is the same for the two theories, while the component normal to $\mathbf{m}$, denoted $\dot{\boldsymbol{\epsilon}}^T$, is found to vanish. As expected, the two types of theory therefore predict the same incremental response, governed by the modulus h, along the fundamental proportional loading path.

To examine the stability of this fundamental solution at a given load level, a small perturbation is now superimposed on the next increment $\dot{\alpha}\boldsymbol{\sigma}^o$ of proportional loading, consisting of a small change in direction that is denoted by $\dot{\boldsymbol{\sigma}}^1$. The strain rate produced by the altered stress rate is still the same in the direction of $\mathbf{m}$, but it is seen from (17) that the deformation theory now predicts a non-vanishing component $\dot{\boldsymbol{\epsilon}}^T = 1/h_1\mathbf{M}{:}\dot{\boldsymbol{\sigma}}^1$ which is absent from the flow theory. This is illustrated in Figure 3b, where the directions of straining normal and tangential to the flow potential are indicated. The flow theory is found to predict an infinitely stiff response in a direction perpendicular to the loading direction, whereas the deformation theory implies a finite stiffness which is determined by the secant modulus h_1, the latter being proportional to the equivalent stress. Consequently, the initiation of a new mode of deformation, which is characteristic for the onset of an instability, is more easily accommodated by a deformation-theory type response than it would be according to flow theory. This difference tends to play a crucial role in most stability analyses and notably also for the results obtained in the next section[3].

It remains now to motivate physically the necessity of having a finite stiffness in the direction perpendicular to the direction of loading, a feature of the deformation theory presented.

3.4 On the link between deformation and corner theories

The presence of a finite stiffness in the direction perpendicular to the proportional loading path is motivated by appealing to the microstructural deformation mechanisms of the rock. These frictional mechanisms are described in a phenomenological way with a flow theory of plasticity in which the yield surface possesses the vertex structure proposed by Koiter (1953).

This yield surface in stress space ϕ is now assumed to be composed of a finite number of regular yield surfaces $\phi^r (r = 1, ...)$. Each surface, representing a microstructural deformation mechanism, is activated independently of the others, if the value of ϕ^r is zero and a loading condition identical to (9) is respected.

[3] In fact, the only counter-example known to the authors is the puckering instability during the forming process of an hemispherical cup from an initially flat plate (Triantafyllidis, 1985). It turns out, however, that the loading along the axisymmetric fundamental solution of that problem is non-proportional, in violation of the basic supposition that explains the destabilizing element in a deformation theory.

The first activation of a surface signals macroscopic yield. If several surfaces are participating to the deformation ($r = 1$ to n) then the total strain rate is build from the contribution of each system composed of the yield surface ϕ^r and flow potential ψ^r. The total strain rate is thus expressed as the following sum:

$$\dot{\boldsymbol{\epsilon}} = \sum_{r=1}^{n} \frac{1}{h^r} \mathbf{m}^r \otimes \mathbf{n}^r : \dot{\boldsymbol{\sigma}} , \tag{18}$$

in which $\mathbf{n}^r$ and $\mathbf{m}^r$ are the normal in stress space to ϕ^r and ψ^r, respectively. The modulus h^r is the work-hardening modulus associated with the plastic flow of system r.

The geometrical interpretation of the construction (18) is illustrated in Figure 3c for the case of two active systems represented by straight lines. The stress rate $\dot{\boldsymbol{\sigma}}$ activates the two yield criteria since its orientation falls within the cone, marked by the dotted lines, in which the loading condition (9), $\mathbf{n}^r : \dot{\boldsymbol{\sigma}} > 0$ for r equal to 1 and 2, is fulfilled. Varying the orientation of that rate $\dot{\boldsymbol{\sigma}}$ within the confine of that cone results in a strain rate which varies in orientation from $\mathbf{m}^1$ to $\mathbf{m}^2$. Indeed, $\dot{\boldsymbol{\epsilon}}$ must be a positive linear combination of these two vectors according to (18). This property to change the orientation of the strain increment with the orientation of the stress increment is shared with the deformation theory presented above but not by the flow version.

Further physical interpretation of this plasticity model with a yield vertex is now in order. This model could be constructed from laboratory measurements on a representative volume element which contains a family of pervasive small faults or cracks with random orientation. Every crack, labeled r, could be activated according to the yield criterion ϕ_r and could participate to the macroscopic deformation according to the flow rule ψ_r, independently of other cracks if interactions are disregarded. A simple procedure was proposed by Batdorf & Budiansky (1949) to compute the macroscale deformation in a way similar to (18) except for the infinite number of systems they considered in an effort to study the response of a representative volume element composed of randomly oriented crystals. It is that idea, applied to fissured rocks, which motivated Rudnicki & Rice (1975) to propose a deformation-theory version of their plasticity model. Leroy & Sassi (this Volume) pursue this idea further with the aim to develop a plasticity model, based on phenomenological laws for discontinua and appropriate for modeling geological deformations at the tectonic scale. The challenge remains to construct a plasticity model from a fine description of the micromechanism as it was considered by Lehner & Kachanov (1995).

4 Application

This section will illustrate the use of solutions to the above variational problem in predicting folding. It will also deal with the difficulty of reproducing field observations in laboratory models, arising mainly from a lack of similarity in the deformation responses of the natural rock and its laboratory analogue material.

We begin with a characterization of a 'prototype' that represents an interpretation and idealization of a geological field example. This is followed by some stability predictions for the prototype. Thereafter, the construction of a physical model of the prototype and, in particular, the selection of appropriate analogue materials are discussed. Finally, our stability predictions are applied to an incompletely scaled model and compared with those obtained for the prototype.

4.1 Field example and stability predictions

As a field example, we have selected a seismic section from a paper by Cobbold & Szatmari (1991) through the Campos salt basin, offshore Brazil (see line E for their Fig. 3 for the section). Here, the gliding of sediments in a down-slope direction is thought to have resulted in a compression of the selected 45 km-long section on the lower slope. This example has been chosen, because the two visible folding events during the Albian and the Tertiary, occurred without localized faulting as far as can be judged from the seismic profile. The first folding event affected an overburden of some 250 m thickness with a wavelength of 2.5 km. The second episode of folding involved the entire overburden, then approximately 2.5 km thick, with a wavelength of 30 km. These numbers obviously represent a simplification of the actual situation. We shall add as a further approximation the assumption of a uniform salt layer thickness of 600 m at all times. The three-layer sequence overburden/substratum/basement so defined constitutes the plane 'prototype' section to be studied. Note that it is assumed that the substratum to basement boundary is horizontal (Demercian *et al.*, 1993) and that the condition of slip with no friction prevails at the basement contact.

The prototype geometry thus coincides with the model configuration of Figure 1, the stability of which has been studied previously by Triantafyllidis & Leroy (1997) with a view on explaining the appearance of the two distinct folding episodes in the presence of an in-situ stress gradient. Their analysis gave an indication of the conditions under which the Albian mode would become dominant. It also provided an explanation for the arrest of that mode and for the triggering of the long-wavelength instability during the Tertiary. In the process, it was shown that a deformation theory of plasticity was necessary to explain folding during the Albian. Moreover, it was shown that differential sedimentation during the Tertiary (modeled here by a redistribution condition) had to complement the destabilizing action of the tectonic stress and of buoyancy, to make possible the initiation of the long-wavelength fold. These results will be discussed presently in greater detail, based on a finite-element solution to the variational problem. First, however, the material data set used in the analysis will be provided.

A short description of the pertinent sedimentary record in the Campos basin may be found in Demercian *et al.* (1993). The complete set of rocks properties, as used and justified by Triantafyllidis & Leroy (1997), is now summarized after a word of caution. Obtaining precise values for the material properties (or, even more so, for their spatial distribution) of the sedimentary rocks in the Campos basin is beyond the scope of this paper. The parameter values used are nevertheless thought to provide reasonable order-of-magnitude estimates that serve to illustrate the potential of our method for explaining the absence of localized faulting as well as the observed folding events. With these estimates, it is already possible to show how the magnitudes of the stress necessary for instability depends on the model chosen (flow or deformation theory, redistribution or not). These magnitudes differ sufficiently for our conclusions to become insensitive to the exact values of the material parameters. The stress parameterization (1) requires the orientation angle φ and the scalars k_1 and k_3 which are set to $\pi/8$, 1, and 1, respectively. The two material densities are chosen to be 2100 and 2200 kg/m^3 for the overburden and substratum, respectively. An elastic modulus of 5 GPa is proposed for the sediments during the Albian whereas a value 10 seems to be more appropriate for the whole sequence during the second stage of folding. Poisson's ratio is set to 0.2. The hardening exponent m is 4 and the yield stress τ_y is set to 10 MPa. The friction and dilatation coefficients μ and β are given the values 0.6 and 0, respectively.

The predictions for the onset of an instability during the Albian are presented in Figure 4 in which isocontours of stability exponent Λ, scaled by the modulus of elasticity in shear G, are plotted in a plane spanned by the dimensionless wavenumber ωH_a and the stress σ_0 normalized by the cohesion τ_y. Solid and dotted curves correspond to the predictions in the absence and in the presence of the redistribution condition at the top surface.

The neutral stability curves ($\Lambda = 0$) partition the graph in stable and unstable regions marked by a minus and plus sign, respectively. The shape of the isocontours obtained with no redistribution are reminiscent of those found for the folding of an elastic plate over a viscous substratum. This shape is close to a parabola for large value of ωH_a signalling the dominance of the classical Euler solution for short wavelength modes. The presence of the salt layer penalizes the development of long-wavelength modes as suggested by the hyperbolic structure of the isocontours in the vicinity of the horizontal axis.

These results, which are thus predicting a folding instability, were obtained for a constant thickness of the overburden corresponding to the 250 m estimated by Demercian *et al.* (1993). They are now analyzed to find the stress conditions necessary to render the dimensionless wavenumber ωH_a dominant at precisely 0.63 (the instability wavelength is 2.5 km). For given stress conditions, the dominant perturbation is defined as the one with fastest growth, thus with the largest stability exponent Λ/G. For example, for a τ_y of -4.3 in Figure 4, there is an isocontour which has an infinite slope at that particular stress. The ωH_a coordinate of that point of infinite slope is the desired 0.63 and defines the dominant perturbation for which the rate of growth is maximum and found to be 1.9×10^{-3}. Note that the conditions for the onset of localized faulting, marked by diamonds in Figure 4 and in the other figures of this paper, require a stress equal to -5.2 τ_y. The prediction is thus that folding could occur without any localized faulting, in agreement with the field observations. The two stresses discussed so far are close to the value of -6 τ_y (-60 MPa) which must have prevailed in the field according

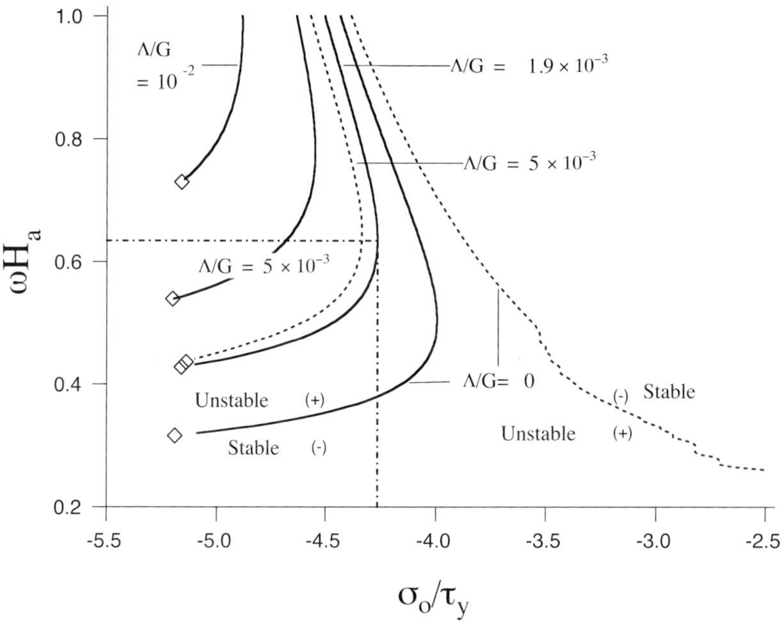

Fig. 4. Results of a stability analysis, giving conditions for the onset of folding and localized faulting for a section of the Campos salient during the Albian. Plotted are a dimensionless perturbation wavenumber versus a normalized stress at the overburden/substratum boundary. The solid (dotted) lines have been obtained with the deformation-theory version of a Rudnicki-Rice model, without (with) a redistribution condition at the top surface. Diamonds marks the onset of localized faulting.

to Triantafyllidis and Leroy (1997). It should be stressed that those results were obtained for the deformation-theory version of Rudnicki and Rice's plasticity model. The use of the flow-version of the same model results in a prediction for the stress necessary to destabilize the section 3 times larger than the magnitude indicated above and is not acceptable.

Figure 4 contains also results obtained in the presence of redistribution at the top surface. The shape of the isocontours, different from the one discussed above for long wavelength (small ωH_a), deserve some comments. The neutral stability isocontour does not have a point of infinite slope and enters the regions of elastic behavior marked by a slope discontinuity on Figure 4. That curve continues on the right of the graph and enters the region of extension. This structure indicates that shallow overburden can be destabilized regardless of the magnitude of the tectonic force. However, for the particular wavenumber 0.63 of interest, the redistribution condition does not influence the shape of the critical isocontour nor the value of the critical stress required. Its influence is limited to the initial rate of growth of the fold which is found to be increased by a factor 2.5, approximately, in the presence of redistribution.

The second aim of this subsection is to explain the arrest of the Albian fold development and to motivate the initiation of a long-wavelength fold during the Tertiary. For that purpose, the observer keeps the wavelength ($2\pi/\omega$) of the instability constant and search for the thickness H_a required to have neutral stability. The assumptions are that the dominant wavelength initially selected is the only one to develop, that new sediments increase the stiffness of the structure and have thus a stabilizing influence, and that the presence of finite-amplitude perturbations does not invalidate the verdict of the linear stability analysis. Furthermore, to explain the fold development arrest, we do not invoke the restabilization of the initially unstable system in a fold of finite amplitude, found in the non-linear stability analyses of Massin *et al.* (1996). From the neutral stability curves in Figure 5, it is seen that the thickness of the overburden has to be close to 290 and 320 m for the instability to cease. These two numbers are obtained in the absence and the presence of redistribution, respectively. They both indicate that the fold development was arrested as the overburden thickened.

The onset of the 30 km-long wavelength mode in the Tertiary remains to be discussed now. The dashed neutral stability curve in Figure 5 shows that, for stress conditions that differ from those in the Albian only by an additional overburden, it is for thicknesses less than 3 km that an instability with a wavelength of 30 km can be initiated. However, these results were obtained assuming a redistribution condition at the top surface. Failure to account for that condition results in a 40 times larger critical stress that is also accompanied by localized faulting. The main difference between the two folding events is thus that a redistribution of sediment was required for triggering the long-wavelength instability.

4.2 Remarks on the Design of a Scaled Model

We now turn our attention to the design of a scaled physical model of the prototype for a laboratory investigation of overburden folding and faulting. A

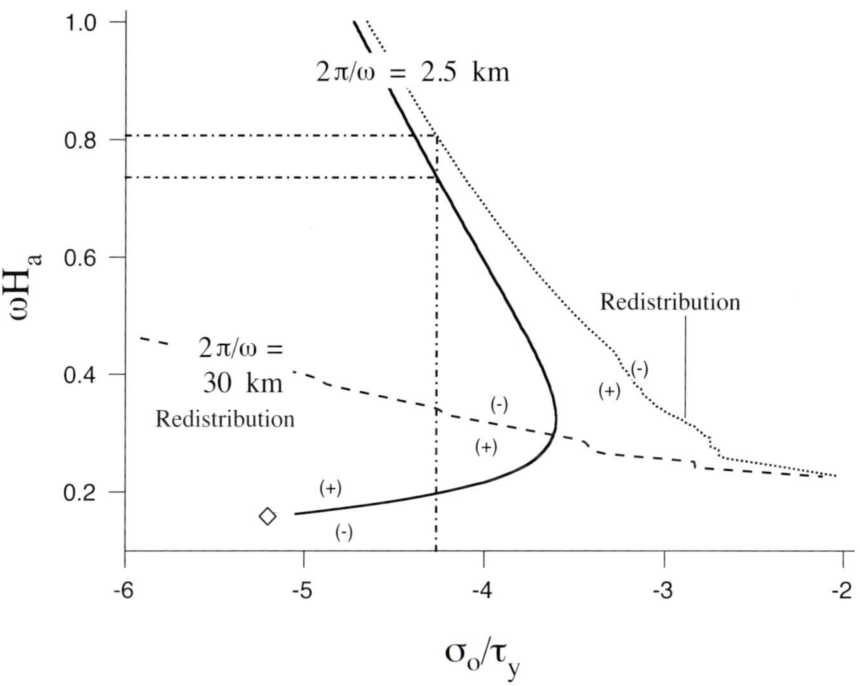

Fig. 5. Neutral stability predictions for constant perturbation wavelengths. Solid and dotted line are for the Albian without and with redistribution condition, respectively. The dashed line is for the Tertiary and is obtained accounting for redistribution.

sand-box experiment is imagined with two layers on a rigid, flat basement. A layer of silicone putty is laid on that basement with the appropriate lubrication of the interface to reproduce the slip with no friction that must prevail in the Campos salient to allow continuous contraction. The viscous material is covered with a dry pure quartz sand such as the one used by Vendeville and Jackson (1992) to mimic the brittle but competent overburden.

To determine the properties and the dimension of the scaled prototype, we follow the steps considered already by Nettleton and Elkins (1947) and invoke the need to satisfy the conditions of similarity. More precisely, it is the conditions of simple similarity - same scaling for displacement and length- which are considered to model appropriately geometric effects as shown by Mandel (1962) in an authoritative essay on similarity rules for nonlinear materials. Consider the material density and length in the analogue model to be those of the field structure multiplied by a factor ρ^* and l^*, respectively, and the gravity acceleration to be preserved. The equations of equilibrium dictates that stress and thus any material parameters having similar dimension, such as the elastic modulus E and the cohesion, be scaled by $g\rho^* l^*$. As pointed out by Mandel, the nonlinear stress-strain curve depicted in Figure 2 must also be the same for the two materials requiring, for the parameterization adopted here, that the hardening exponent m be the same. The scaling of the only time scale of the problem, the relaxation time of the viscous layer, is not constrained in the absence of inertia forces.

The parameter selection for the scaled prototype starts with the recognition that the smallest length of the structure (250 m) must be scaled down at most to a centimeter if sand grain of the size 300 μm are used. The total length of the sand box is thus of the order of 2 m to scale the 45 km of the field structure and the scalar l^* has for value 4×10^{-5}. The dry sand has a cohesion of 300 Pa ($c^* = 3.0 \times 10^{-5}$), a friction coefficient of 0.6 (Krantz, 1991) and a density of $1760\,\mathrm{kg/m^3}$ ($\rho^* = 0.76$) (Vendeville and Cobbold, 1987). Its elastic properties are rarely mentioned in the literature and it is assumed, from the results of wave propagation experiments in sand under confining pressure of a few Pa, that the elastic modulus is of the order of 10^9 Pa (Nieuwland, 1994). Poisson's ratio is set to a realistic 0.2. The viscous material is similar to the one considered by Vendeville and Jackson (1992) and its density adjusted, thanks to the addition of minerals, to the ideal value of 1680 kg/m^3. Its viscosity is assumed constant for all strain rates but its exact value is not required in our analysis. From the choice of length scale in the laboratory and of the analogue material density it appears that the cohesion scaling factor c^* is close to the ideal value ($g\rho^* l^* = 2.9 \times 10^{-4}$). However, it is known that the rule of similarity is far from being satisfied by the elastic properties of the analogue material which has an elastic modulus 4 orders of magnitude larger than the ideal one.

4.3 Folding and Localized Faulting in the Laboratory

The stability predictions are now proposed for the scaled model in which the 1 cm thick overburden (250 m in the field during the Albian) is composed of the quartz sand described above. It is shown that it is necessary to replace this quartz sand by an ideal analogue material to reduce the elastic stiffness and to model correctly the initiation of folding. The importance of the deformation mechanism in the overburden and the potential error introduced by the analogue model, if that mechanism is not duplicated appropriately in the laboratory, is also discussed.

The results for the instability of the scaled physical model with quartz sand could not be presented on the same graph with those for the prototype of the Albian layer because the stress required to fold the analogue model was 20 times larger than the ones predicted for the real structure and always associated with localized faulting. As already mentioned during the presentation of the scaled model, the elasticity modulus of quartz sand does not scale properly and folding of the structure cannot be predicted adequately. It is now proposed to replace the quartz sand by an ideal granular material with an elastic modulus which has exactly the theoretical value $g\rho^* l^* E$ found by the scaling rule. The rest of its mechanical properties are identical to the ones reported for the quartz sand. The results of the predictions are compared to those obtained for the Campos salient in Figure 6a. The difference in terms of stress and conditions for the initiation of localized faulting is small and acceptable: folding and faulting will be duplicated in the laboratory appropriately.

However, the new, yet imaginary, analogue material for the overburden could have the appropriate linear and nonlinear responses in a laboratory test but could differ from the quartz sand by its way to accommodate deformation at the micro scale. Sand is known to deform by isolated failure prior to maximum load (Arthur *et al.*, 1977). Theses events have been modeled by slip across heterogeneities (Shi and Horii, 1989) and could be well represented by the deformation theory adopted for our predictions. If a similar mechanism is not present in the ideal analogue material, the use of a flow theory of plasticity is then more appropriate. In that instance, the stability predictions are presented in Figure 6b and compared with those obtained with the deformation theory. It is observed that the difference in compressive stress and in dominant wavelength is about a factor 2 between the two sets of predictions. This result illustrates the sensitivity of the stability predictions to the detail of the constitutive model and shows the difficulty to scale stress at the onset of localized faulting and of folding from the laboratory to the field.

5 Conclusion

The first of the three objectives of this paper is to show the potential of elasto-plasticity constitutive models for studying simultaneously the onset of two different modes of instability which are localized faulting and folding. The second aim is to present the sensitivity of the stability predictions to the details of the constitutive model adopted. Emphasis is put on the use of a deformation theory of plasticity which, in linear stability or bifurcation analyses, mimics the presence of a population of pervasive faults at the micro-scale. A detailed justification

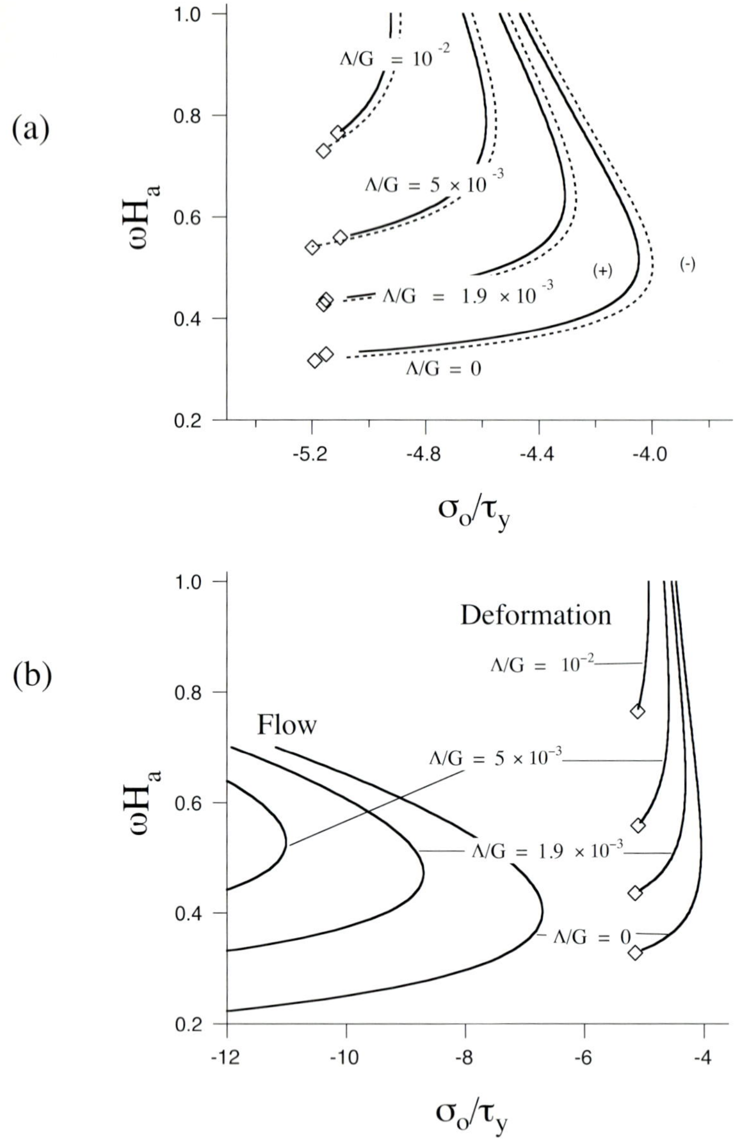

Fig. 6. Comparison of stability predictions for the Albian folding between scaled model and prototype. In (a), the solid lines are for the prototype and the dashed curves correspond to the scaled model built with an analogue material which has the ideal elastic properties. In (b), results are presented for the scaled prototype with an analogue material deforming according to either the flow- or the deformation-theory version of Rudnicki and Rice's plasticity model. Diamonds mark the onset of localized faulting.

of that analogy is provided as well as a complete derivation of the incremental moduli necessary for our stability analysis. The third objective of this paper is to show the difficulty in reproducing in the laboratory the instability modes detected in the field.

The solution to the stability variational problem is applied to a radial section through the Campos salient, offshore Brazil, reported by Cobbold and Szatmari (1991) and Demercian et al. (1993). The two folding events that occurred during the Albian and the Tertiary, in the absence of localized faulting, are explained. Despite the difficulty in selecting material parameters, it is shown that a deformation theory of plasticity is required to predict an instability on the appropriate wavelength and for realistic stress conditions in the Albian. Furthermore, it is shown that the destabilizing action of the tectonic force and of the density contrast has to be complemented by differential sedimentation (modeled as a redistribution condition) to explain the long-wavelength folding during the Tertiary.

These predictions are the results of linearized calculations and need to be verified with nonlinear analyses often conducted by numerical means. For example, the finite-element analyses of Massin et al. (1996) have shown that an initially unstable stratified structure, such as the one studied here, will evolve in time to take the shape of a stable fold of finite amplitude. That amplitude is a function of the initial tectonic stress considered. The gradient with depth of that in-situ stress, studied by Sassi and Faure (1997), appears to be more complex than assumed here if layer parallel slip is accounted for. The presence of localized faulting within the overburden can only add to that complexity (Poliakov and al., 1993). In all those numerical studies, a constitutive model analogous to a flow theory of plasticity has been used which, from our linear stability results, may not be appropriate for studies at the tectonic scale. The development of constitutive relations from a description of the dominant dissipative micro-mechanism in the field is a necessary step to be considered in the future.

This research direction has been identified in the past for engineering applications and only a few of the numerous publications found in the literature are commented here. The macroscopic yield surfaces for heterogeneous materials are expected to exhibit a yield vertex (Hill, 1967) as already seen in the early work of Batdorf and Budiansky (1947). For rocks, Lehner and Kachanov (1995) have shown that the sliding along preferentially oriented cracks in a population of randomly oriented, non-interacting pervasive faults results indeed in the formation of a vertex in the macroscopic yield surface at the loading point. Furthermore, the deformation at the current loading point is path independent in a restricted region in stress space. Crack interaction, either accounted for approximately (Costin, 1985) or by numerical means (Horii and Nemat Nasser, 1983) results in variation in the description of that vertex but confirms its existence (Rudnicki and Chau, 1996). The presence of a corner or a reduction of curvature on the yield surface at the loading point is corroborated from the results of combined compression-torsion tests on Tennesse marble (Olsson, 1992). From these references, it is clear that constitutive relations with the appropriate features for analysis at the tectonic scale should become available in the future once the questions of choice of micro-mechanism and of size of representative volume has been discussed. They will improve our modeling capabilities, compared to classical models found in the geotechnical literature, and will provide the means to relate bulk deformation and anisotropic development of pervasive fracturing. That link is important to model fractured reservoir as it is discussed by Leroy and Sassi in this volume.

References

Arthur, J.R.F., Dunstan, T., Al-Ani, Q.A.J. and Assadi, A. (1977). Plastic deformation and failure in granular media. *Géotechnique*, 27, 53–74.

Batdorf, S.B. and Budiansky, B. (1949). A mathematical theory of plasticity based on the concept of slip. *U.S. N.A.C.A. Technical Note 1871.*

Biot, M.A. (1961). Theory of folding of stratified visco-elastic media and its implication in tectonics and orogenesis. *Geol. Soc. Am. Bull.*, 72, 1595–1620.

Biot, M.A. and Odé, H. (1965). Theory of gravity instability with variable overburden and compaction. *Geophysics*, 30, 213–227.

Breckels, I.M. and van Eekelen, H.A.M. (1982). Relationship between horizontal stress and depth in sedimentary basins, *J. Pet. Tech.*, 2191–2199.

Budiansky, B. (1959). A reassessment of deformation theories of plasticity. *J. of Applied Mechanics*, 259–264.

Cobbold, P.R. and Szatmari, P. (1991). Radial gravitational gliding on passive margins. *Tectonophysics*, 188, 249–289.

Costin, L.S. (1985). Damage mechanics in the post-failure regime, *Mechanics of Materials*, 4, 149–160.

Demercian, S., Szatmari, P. and Cobbold, P.R. (1993). Style and pattern of salt diapirs due to thin-skinned gravitational gliding, Campos and Santos basins, offshore Brazil. *Tectonophysics*, 228, 393–433.

Drucker, D.C. and Prager, W. (1952). Soil Mechanics and plastic analysis or limit design. *Q. J. Appl. Math.*, 10, 157–165.

Hill, R. (1950). *The Mathematical Theory of Plasticity*, Clarendon Press, Oxford.

Hill, R. (1958). A general theory of uniqueness and stability in elasic-plastic solids. *J. Mech. Phys. Solids*, 6, 236–249.

Hill, R. (1967). The essential structure of constitutive laws for metal composites and polycrystals. *J. Mech. Phys. Solids*, 15, 79–95.

Horii, H. and Nemat-Nasser, S. (1983). Estimate of stress intensity factors for interacting cracks. In: *Advances in Aerospace Structures, Material and Dynamics*, 111–117, edited by U. Yuceoglu, R.L. Sierakowski and D.A. Glasgow, ASME, New-York.

Hutchinson, J.W. (1970). Elastic-plastic behaviour of polycrystalline metals and composites. *Proc. Roy. Soc. Lond.*, A-319, 247–272.

Hutchinson, J.W. (1974). Plastic Buckling. *Adv. Appl. Mech.*, 14, pp. 67–144.

Koiter, W.T. (1953). Stress-strain relations, uniqueness and variational theorem for elastic-plastic materials with a singular yield surface. *Q. Appl. Math.*, 11, 350–354.

Kötter, F. (1903). Die Bestimmung des Druckes an Gekrümmten Gleitflächen, *Ber. Akad. der Wiss.*, Berlin.

Lehner, F.K. and Kachanov, M.L. (1995). On the stress-strain relations for cracked elastic materials in compression. In: *Mechanics of Jointed and Faulted Rock*, 49–61 edited by H.P. Rossmanith, Balkema, Rotterdam.

Leroy, Y.M. and Triantafyllidis, N. (1996). Stability of a frictional, cohesive layer on a substratum: variational formulation and asymptotic solution. *J. Geophys. Res.*, 101, *B8*, 17795–17811.

Leroy, Y.M. and Triantafyllidis, N. (1998). Stability of layered geological structures: an asymptotic solution. In: *Material Instabilities in Solids*, 15–25, edited by E. van der Giessen and R. de Borst, J. Wiley & Sons, New York.

Leroy, Y.M. and Sassi, W. (1999). A plasticity model for discontinua. This Volume.

Mandel, J. (1962). Essais sur modèles réduits en mécanique des terrains. Etude des conditions de similitude. *Rev. de l'Industrie Minérale*, 44, 611–620. (English translation in *Int. J. Rock Mech. Mining Sci.*, 1, 31–42, 1963.)

Mandl, G. (1988). Mechanics of tectonic faulting. Elsevier, Amsterdam.

Mandl, G. and Fernandez Luque, R. (1970). Fully developed plastic shear flow of granular materials. *Géotechnique*, 20, 277–307.

Massin, P., Triantafyllidis, N. and Leroy, Y.M. (1996). Stability of density-stratified two-layer system. *C. R. Acad. Sci., Sér. IIa, Tectonique*, 322, 407–413.

McGarr, A. and Gay, N.C. (1978). State of stress in the Earth's crust. *Ann. Rev. Earth Planet. Sci.*, 6, 405–436.

Neale, K.W. (1981). Phenomenological constitutive laws in finite plasticity. *Solid Mechanics Archive*, 6, 79–128.

Nettleton, L.L. and Elkins, A. (1947). Geologic models made from granular materials. *Trans. Am. Geophys. Union*, 28, 451–466.

Nieuwland, D.A. (1994). Personal communication.

Odé, H. (1960). Faulting as a velocity discontinuity in plastic deformation. In: *Rock Deformation*, edited by D. Griggs and J. Handin, Geol. Soc. Am. Memoir 79, pp. 293–321.

Ogden, R.W. (1984). *Non-Linear Elastic Deformations*, Ellis Horwood, Chichester, England.

Olsen, W.A. (1992). The formation of yield-surface vertex in rock. In: *Proc. 33rd U.S. Symposium Rock Mech.*, pp. 701–705, Balkema, Rotterdam.

Poirier, J.P. (1980). Shear localization and shear instability in materials in the ductile field. *J. Structural Geology*, 2, 135–142.

Poliakov, A.N.B., Podladchikov, Y. and Talbot, C. (1993). Initiation of salt diapirs with frictional overburdens: numerical experiments. *Tectonophysics*, 228, 199–210.

Ramberg, H.R. and Stephansson, O. (1964). Compression of floating elastic and viscous plates affected by gravity, a basis for discussing crustal buckling. *Tectonophysics*, 1, 101–120.

Rice, J.R. (1976). The localization of plastic deformation. In: *Theoretical and Applied Mechanics*, Proc. of the 14th IUTAM Conference, edited by W. Koiter, pp. 207–220, North Holland, Amsterdam.

Rudnicki, J.W. (1984). A class of elastic-plastic constitutive laws for brittle rock. *J. of Rheology*, 28, 759–778.

Rudnicki, J.W. and Rice, J.R. (1975). Conditions for the localization of the deformation in pressure-sensitive dilatant materials. *J. Mech. Phys. Solids*, 23, 371–394.

Rudnicki, J.W. and Chau, K.T. (1996). Multiaxial response of microcrack constitutive model for brittle rock. In: *NARMS'96*, 2, edited by M. Aubertin, F. Hassani and H. Mitri, pp. 1707–1714, Balkema, Rotterdam.

Sanders, J.L. (1954). Plastic stress-strain relations based on linear loading functions. In: *Proc. Second US Nat. Cong. Appl. Mech.*, edited by P.M. Naghdi, ASME, 455–460.

Sassi, W. and Faure, J.L. (1996). Role of faults and layer interfaces on the spatial variation of stress regime in basins: Inference from numerical modelling. *Tectonophysics*, in press.

Shanley, F.R. (1947). Inelastic column theory. *J. of the Aeronautical Sciences*, 14, 261–267.

Shi, Z.H. and Horii, H. (1989). Microslip model of strain localization in sand deformation. *Mechanics of Materials*, 8, 89–102.

Smoluchowski, M. (1909). Über ein gewisses Stabilitätsproblem der Elastizitätslehre und dessen Beziehung zur Entstehung von Faltengebirgen. *Abhandl. Akad. Wiss Krakau, Math. Kl.*, 3–20.

Sokolovski, V.V. (1960). Statics of soil media. Butterworth, London.

Spencer, A.J.M. (1964). A theory of the kinematics of ideal soils under plane strain conditions. *J. Mech. Phys. Solids*, 12, 337–351.

Spencer, A.J.M. (1971). Discussion of "Fully developed shar flow of granular materials, by G. Mandl and R. Fernandez-Luque". *Géotechnique*, 21, 190–192.

Spencer, A.J.M. (1982). Deformation of ideal granular materials. In : H.G. Hopkins and M.J. Sewell (editors), *Mechanics of Solids*, Pergamon.

Stören, S. and Rice, J.R. (1975). Localized necking in thin sheets. *J. Mech. Phys. Solids*, 23, 421–441.

Triantafyllidis, N., (1985). Puckering instability phenomena in the hemispherical cup test. *J. Mech. Phys. Solids*, 33, 117–139.

Triantafyllidis, N., Needleman, A. and Tvergaard, V. (1982). On the development of shear bands in pure banding. *Int. J. Solids and Structures*, 18, 121–138.

Triantafyllidis, N. and Lehner, F. K. (1993). Interfacial instability of density-stratified two-layer systems under initial stress. *J. Mech. Phys. Solids*, 41, 117–142.

Triantafyllidis, N. and Leroy, Y.M. (1994). Stability of a frictional material layer resting on a viscous half-space. *J. Mech. Phys. Solids*, 42, 51–110.

Triantafyllidis, N. and Leroy, Y.M. (1997). Stability of a frictional, cohesive layer on a substratum: validity of asymptotic solution and influence of material properties, *J. Geophys. Res.*, 102, *B9*, 20551–20570.

Tvergaard, V., Needleman, A. and Lo, K.K. (1981). Flow localization in the plane strain tensile test. *J. Mech. Phys. Solids*, 2, 115–142.

Vendeville, B. and Cobbold, P.R. (1987). Glissements gravitaire synsédimentaires et failles normales listriques: modéles expérimentaux. *C. R. Acad. Sci. Paris*, 305, Série II, 1313–1319.

Vendeville, B.C. and Jackson, M.P.A. (1992). The rise of diapirs during thin-skinned extension. *Marine and Petroleum Geology*, 9, 331–353.

6 Appendix: Derivation of incremental moduli

The objective of this Appendix is to derive the incremental moduli of the flow and deformation theories of Rudnicki and Rice's model. The argument is similar to the one considered in section 3 and extended here to account for the elastic compliance of the material disregarded so far. The extension of the incremental relations to the large deformation context appropriate for stability or bifurcation analyses is also discussed.

The incremental relation between stress and strain for the flow theory reads:

$$\dot{\boldsymbol{\sigma}} = \mathcal{L}^e : (\dot{\boldsymbol{\epsilon}} - \dot{\gamma}\mathbf{m}) , \tag{19}$$

where the fourth-order tensor $\mathcal{L}^e$ for an isotropic elastic solid takes the form:

$$\mathcal{L}^e = 2G\mathbb{I} + (K - \frac{2G}{3})\mathbf{I} \otimes \mathbf{I}, \tag{20}$$

in which K is the modulus of elasticity for bulk deformation. The decomposition of the strain rate in an elastic and a plastic contribution as well as the Prandtl Reuss-type of flow law (6) characteristic of flow theories are adopted to derive (19). The flow potential chosen is the one presented in (11) with a constant dilatancy coefficient and a normal in stress space $\mathbf{m}$ defined in (12). The yield criterion is also the one considered in section 3 and the consistency condition on that criterion provides an expression for $\dot{\gamma}$ in terms of the strain rate by combining (10) and (19):

$$\dot{\gamma} = \frac{1}{H}\mathbf{n} : \mathcal{L}^e : \dot{\boldsymbol{\epsilon}}, \tag{21}$$
$$\text{with} \quad H = h + \mathbf{n} : \mathcal{L}^e : \mathbf{m}.$$

The moduli for the flow theory are now obtained by combining this last result with relation (19) to eliminate the equivalent strain rate:

$$\dot{\boldsymbol{\sigma}} = \mathcal{L}_F^{ep} : \dot{\boldsymbol{\epsilon}}, \tag{22}$$
$$\text{with} \quad \mathcal{L}_F^{ep} = \mathcal{L}^e - \frac{1}{H}\mathcal{L}^e : \mathbf{m} \otimes \mathbf{n} : \mathcal{L}^e.$$

This expression, required for our stability analysis, is valid under the condition of loading which is expressed by: $\dot{\phi} = 0$ and the inequality (9).

To derive the moduli for the deformation theory, it is found convenient to first decompose the incremental response of the flow theory in a spherical and deviatory part making use of the definition of the isotropic elasticity tensor (20):

$$\dot{\boldsymbol{\epsilon}}' = \frac{\dot{\boldsymbol{\sigma}}'}{2G} + \frac{1}{h}\mathbf{m}' \otimes \mathbf{n} : \dot{\boldsymbol{\sigma}}, \tag{23}$$
$$\mathbf{I} : \dot{\boldsymbol{\epsilon}} = \frac{\mathbf{I} : \dot{\boldsymbol{\sigma}}}{3K} + \frac{1}{h}\mathbf{I} : \mathbf{m} \otimes \mathbf{n} : \dot{\boldsymbol{\sigma}},$$

in which the equivalent strain rate $\dot{\gamma}$ has been replaced by its expression in terms of the stress rate (10) and not in terms of the strain rate as above in (21). The hypothesis of proportional loading is now invoked and, following the same argument as in section 3, the two relations in (23) are integrated through the loading sequence to obtain:

$$\boldsymbol{\epsilon}' = \frac{\boldsymbol{\sigma}'}{2G} + \mathbf{m}^{o\prime}\gamma, \tag{24}$$
$$\mathbf{I} : \boldsymbol{\epsilon} = \frac{\mathbf{I} : \boldsymbol{\sigma}}{3K} + \mathbf{I} : \mathbf{m}^o\gamma.$$

Note that the consistency condition (10) was used implicitly while obtaining these expressions. The incremental moduli of the deformation theory results from the rate form of relations (24) obtained while relaxing the hypothesis of proportional loading:

$$\dot{\boldsymbol{\epsilon}}' = \frac{\dot{\boldsymbol{\sigma}}'}{2G} + \frac{1}{h}\mathbf{m}' \otimes \mathbf{n} : \dot{\boldsymbol{\sigma}} + \frac{1}{h_1}[\dot{\boldsymbol{\sigma}}'\frac{1}{2} - \frac{\boldsymbol{\sigma}'}{2\tau} \otimes \frac{\boldsymbol{\sigma}'}{2\tau} : \dot{\boldsymbol{\sigma}}'] , \quad (25)$$
$$\mathbf{I} : \dot{\boldsymbol{\epsilon}} = \frac{\mathbf{I} : \dot{\boldsymbol{\sigma}}}{3K} + \frac{1}{h}\mathbf{I} : \mathbf{m} \otimes \mathbf{n} : \dot{\boldsymbol{\sigma}} .$$

The second equation in (25) concerns the volumetric strain rate and is identical to (23_2), presented for the flow theory, because the spherical part of the normal $\mathbf{m}$ is a constant tensor. However, the first equation in (25) concerning the deviatory part of the strain rate differs from the result obtained for the flow theory due to the introduction of the rate $\dot{\mathbf{m}}'$. That rate corresponds to the third term on the right-hand side of this equation which was expressed in terms of stress rate using expression similar to those in (16).

The objective is now to present the moduli for the deformation theory in a way similar to the one adopted for the flow theory in (22). For that purpose, the two equations in (25) should have a structure similar to those in (23). More precisely, the first equation in (25) should be written as:

$$\dot{\boldsymbol{\epsilon}}' = \frac{\dot{\boldsymbol{\sigma}}'}{2\bar{G}} + \frac{1}{\bar{h}}\bar{\mathbf{m}}' \otimes \bar{\mathbf{n}} : \dot{\boldsymbol{\sigma}} . \quad (26)$$

in which the normal $\bar{\mathbf{n}}$ and $\bar{\mathbf{m}}$ differs from $\mathbf{n}$ and $\mathbf{m}$, defined in (9) and (12), only by the introduction of different coefficients of friction and dilatancy $\bar{\mu}$ and $\bar{\beta}$. Comparing equations (26) and (25_1), it is clear that the modified elasticity modulus in shear $\bar{G}$ is defined by :

$$\bar{G}^{-1} = G^{-1} + h_1^{-1} , \quad (27)$$

and further inspection permits to identify the hardening modulus $\bar{h}$ and friction coefficient $\bar{\mu}$ as

$$\bar{h}^{-1} = h^{-1} - h_1^{-1} \quad \text{and} \quad \bar{\mu} = \mu\frac{h_1}{h - h_1} . \quad (28)$$

The second equation in (25) is now rewritten as follows to reflect the introduction of the modified elastic constants, plasticity modulus and flow normal:

$$\mathbf{I} : \dot{\boldsymbol{\epsilon}} = \frac{\mathbf{I} : \dot{\boldsymbol{\sigma}}}{3\bar{K}} + \mathbf{I} : \frac{1}{\bar{h}}\bar{\mathbf{m}} \otimes \bar{\mathbf{n}} : \dot{\boldsymbol{\sigma}} . \quad (29)$$

The comparison of (29) and (24_2) with the use of the definitions (28) and (27) provides the modified bulk modulus of elasticity and the coefficient of dilatancy:

$$\bar{K}^{-1} = K^{-1} + \frac{\mu\beta}{h_1 - h} \quad \text{and} \quad \bar{\beta} = \beta \frac{h_1}{h - h_1} \,. \tag{30}$$

Having obtained a similar structure for the incremental equations of the deformation- and flow-theory versions, we now propose for the moduli of the deformation theory, similarly to (22),

$$\dot{\boldsymbol{\sigma}} = \mathcal{L}_D^{ep} : \dot{\boldsymbol{\epsilon}} \,, \tag{31}$$

$$\text{with} \quad \mathcal{L}_D^{ep} = \bar{\mathcal{L}}^e - \frac{1}{\bar{H}} \bar{\mathcal{L}}^e : \bar{\mathbf{m}} \otimes \bar{\mathbf{n}} : \bar{\mathcal{L}}^e \,.$$

in which the elastic stiffness $\bar{\mathcal{L}}^e$, the normals $\bar{\mathbf{m}}$ and $\bar{\mathbf{n}}$ and the modulus $\bar{H}$ are defined with the help of the constants introduced in (27), (28) and (30).

The extension of these results to the context of large deformation, necessary for stability and bifurcation analyses, is the last issue discussed in this Appendix. The incremental relations considered have to satisfy the objectivity requirement. For that purpose, the Jaumann rate of the Kirchhoff stress defined by

$$\overset{\nabla}{\boldsymbol{\tau}} = \dot{\boldsymbol{\tau}} + \boldsymbol{\tau} \cdot \mathbf{W} + \mathbf{W}^T \cdot \boldsymbol{\tau} \,, \tag{32}$$

in which $\mathbf{W}$ is the antisymmetric part of the velocity gradient, is often proposed. A superscript T in (32) denotes the transpose of the quantity of interest. This stress rate is related to the rate of deformation tensor which is the symmetric part of the velocity gradient by the incremental moduli:

$$\overset{\nabla}{\boldsymbol{\tau}} = \mathcal{L}^{ep} : \mathbf{D} \,, \tag{33}$$

in which the operator $\mathcal{L}^{ep}$ is either the flow or deformation moduli defined in (22) and (31), respectively. In our stability analysis, the reference configuration is chosen to coincide with the current configuration and the moduli $\mathbf{L}$ entering (3) is then expressed by

$$\mathbf{L} = \mathcal{L}^{ep} + \mathbf{G} \,, \tag{34}$$

$$\text{with} \quad G_{ijkl} = \frac{1}{2}\left[-\sigma_{ik}\delta_{jl} - \sigma_{il}\delta_{jk} + \sigma_{lj}\delta_{ik} - \sigma_{kj}\delta_{il}\right] \,,$$

Further discussions on constitutive relations at large strain is found in the review paper by Neale (1981).

Structural evolution within an extruding block: model and application to the Alpine-Pannonian system

Franz Neubauer[1], Harald Fritz[2], Johann Genser[1], Walter Kurz[1], Franz Nemes[1], Eckart Wallbrecher[2], Xianda Wang[1], and Ernst Willingshofer[3]

[1] Institut für Geologie und Paläontologie, University of Salzburg, Hellbrunner Str. 34, A-5020 Salzburg, Austria
[2] Institut für Geologie und Paläontologie, University of Graz, Heinrichstr. 26, A-8010 Graz, Austria
[3] Dept. of Earth Sciences, Vrije Universiteit, De Boelelaan 1081, NL-1085 Amsterdam, Netherlands

Abstract. Continental escape or lateral extrusion often results from late-stage contraction within continental collision zones when convergence is partitioned into orthogonal contraction, crustal thickening, surface uplift, and sideward motion of fault-bounded blocks. Geometrical arguments suggest that each individual fault-bounded block suffers a specific sequence of deformation. The style of deformation also depends on the location within the block. This includes: (1) initial shortening at the continental couple (future zone of maximum shortening: ZMS); (2) formation of a conjugate shear fracture system and initiation of orogen-parallel displacement of the decoupled extruding block away from the ZMS; (3) because of the changing width of the escaping block away from the ZMS the style of internal deformation changes within the extruding block: (i) shortening (thrusting, folding), surface uplift at the ZMS; (ii) strike-slip faulting along confining wrench corridors and formation of pull-apart basins at oversteps of en echelon shear fractures; (iii) extension parallel and perpendicular to the displacement vector far away from the ZMS. (4) Finally, the extruding block is gradually overprinted by general, laterally expanding contraction that starts to develop from the ZMS. This inferred sequence of deformation is tested by the Oligocene to Recent development of the Alpine-Pannonian system where late stage formation and extrusion of an orogen-parallel block started during the Oligocene. Stages 2 and 3 developed during Early to Middle Miocene, and final general contraction occurred during Late Miocene to Recent.

1 Introduction

Lateral extrusion and escape tectonics have played an important role in the explanation of intracontinental deformation during the late stage of continental collision since the initial formulation of this concept and its application to Cenozoic tectonics of eastern Asia (Tapponnier and Molnar, 1976). The basic concepts state that deformation in front of a rigid indenter is taken up by fault-bounded blocks that essentially move about perpendicular to the displacement direction of the rigid indenter (Fig. 1). Fault orientations in front of the rigid indenter are explained by the slip-line theory that, together with results from analogue modelling (Tapponnier et al., 1982; Ratschbacher et al., 1991a), predict that fault arrangments depend on the shape of the rigid indenter, the orientation of the rigid, undeformable front of the indenter, and on the lateral confinement of the extruding fault blocks. As an essential result, each fault-bounded block that is laterally confined by conjugate strike-slip faults moves individually along an independent displacement vector (Fig. 1).

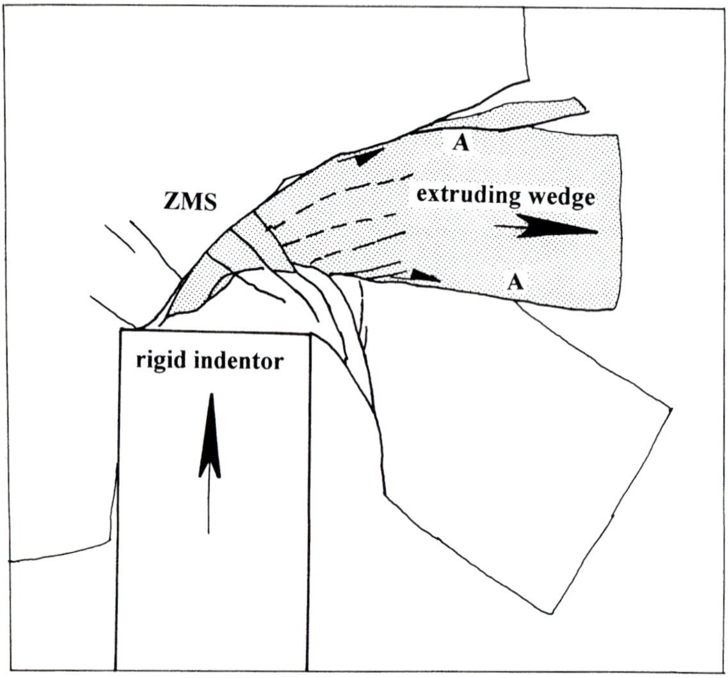

Fig. 1. Sketch drawing of the analogue model displaying formation and lateral displacement of an extruding wedge in front of a rigid indenter (after Tapponnier et al., 1982). Note transtensive sectors of the confining wrench corridors and difference between displacement vector and orientation of confining wrench corridors.

In this contribution we discuss some geometrical effects that arise during the progressive deformational evolution within a single extruding block and confront these theoretical results with a natural example as given in the Cenozoic evolution of the Alpine-Pannonian system.

2 Model of an escaping block

Among the many different shapes and arrangments of fault-bounded extruding wedges we discuss the case of a wedge bounded by conjugate strike-slipe faults that are laterally unconfined, and that moves about perpendicularly to the convergence direction of the indenter as the foreland moves laterally away from the front of the indenter (Figs. 1, 2). In the zone of maximum shortening (ZMS), in front of the indenter, horizontal shortening is partitioned into subvertical thickening and lateral displacement. The ratio between these two components may be variable during progressive evolution. The ZMS is shortened by folding, thrusting, and thickening of the crust as expressed by surface uplift and deepening of the Moho and lateral extrusion. The extruding wedge is laterally confined by conjugate wrench corridors with opposite sense of displacement. The angle between two wrench corridors laterally may vary along the extruding wedge because of secondary shortening within the zone of shortening in front of the indenter. The displacement direction of the extruding wedge may essentially follow the intersection line between the two confining wrench corridors. Consequently, the displacement vector may be largely parallel to sectors of en echelon arranged strike-slip faults within the confining wrench corridor. Furthermore, these sectors may have a major transtensive component during progressive evolution as it is apparent from analogue modelling (A in Fig. 1).

The variable width of the extruding wedge bears some further interesting geometrical implications during progressive displacement of the extruding wedge. The surface topography of the extruding wedge is controlled by the change from shortening in front of the indenter to extension laterally away from the indenter. Furthermore, at a given moment of the development of the extruding wedge, deformation within the extruding wedge changes from shortening, thickening, and surface uplift in front of the indenter through conjugate strike-slip displacement passing the lateral end of the indenter to extension perpendicular to the displacement direction laterally away from the indenter. This implies, furthermore, that during progressive lateral displacement, a zone that was initially shortened passes into a zone of extension perpendicular to the shortening direction (Fig. 2).

The initially laterally unconfined margin of the extruding wedge is transformed into a confined lateral margin, simply because the lateral open realm is filled up. The entire system changes to general contraction, and the zone of shortening may laterally widen along the extruding wedge. This implies that the entire system passes over to general contraction (Fig 3b).

3 Application to the Alpine-West Pannonian system

The structural relationships between the Eastern Alps and the eastwards adjacent Miocene-Pliocene Pannonian Basin suggest geometrical and mass balancing between late orogenic N-S crustal shortening within the Alps, and subsidence within the Pannonian Basin which is related to lateral extrusion (e.g.,

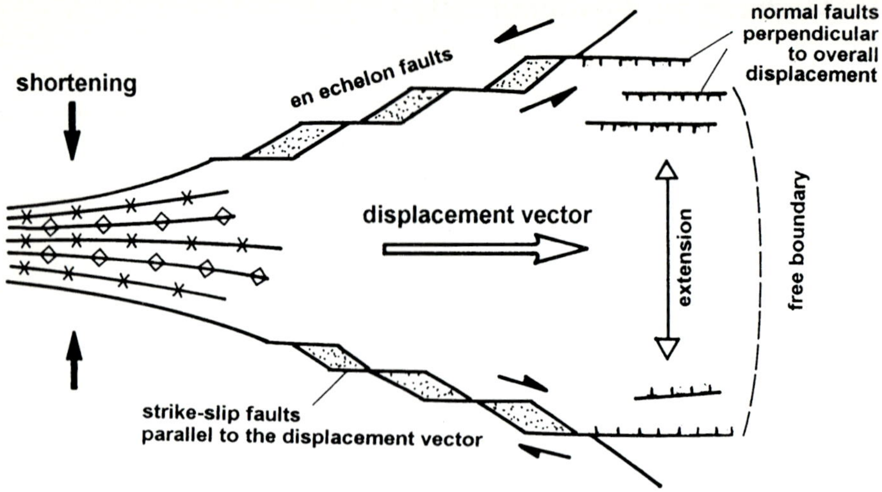

Fig. 2. Predictions of structures that form during extrusion within an individual extruding wedge.

Balla, 1985; Neubauer and Genser, 1990; Ratschbacher et al., 1991b) (Fig. 4a). The confining fault system of the extruding wedge forms a set of conjugate shear zones/strike-slip faults that bound an extruding wedge opening to the east, where the Pannonian basin formed. These confining faults are the sinistral Salzach-Enns fault zone in the north and the dextral Periadriatic fault in the south (Neubauer, 1988; Sprenger and Heinisch, 1988; Ratschbacher et al., 1991b; Nemes et al., 1995; Wang et al., 1996). Accepting these assumptions, a specific N-S section within the escaping wedge carrying the Pannonian Basin should have undergone several stages of deformation during the extrusional process (Fig. 4). These include: (1) maximum N-S shortening within ca. the Tauern window section of the Alps due to the indentation of the Southalpine block into the Alpine system; (2) lateral displacement and formation of pull-apart basins along its margins; (3) N-S extension after escaping the contractional sector and formation of half grabens and extensional basins perpendicular to the ca. E-W displacement vector of the extruding block; and (4) reorganisation during passing the Bohemian indenter that forms an indenter into the Alpine front from the north (Ratschbacher et al., 1991b; Fodor, 1995).

A detailed structural study has been carried out including (i) the structural setting of local, fault-related sedimentary basins on transpressive and transtensive oversteps along the confining wrench systems; (ii) the sequence of deformation events and changes of palaeostress orientation patterns along the margins of and within the extrusional wedge as these can be deduced from mappable faults; and (iii) elucidation of the role of the rheology within various vertical sections of the extruding wedge. A detailed study on all these topics including palaeostress patterns will be published elsewhere.

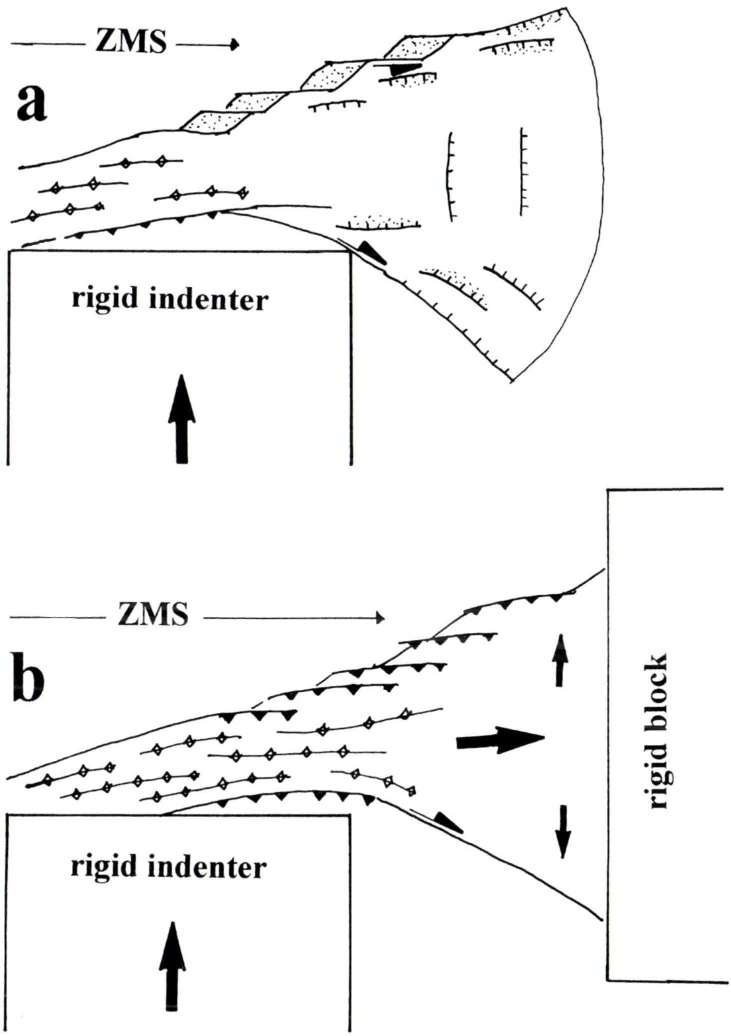

Fig. 3. Predictions of structures that develop during progressive evolution of an extruding wedge. a - Extrusional stage. - b - Blocking of the system by widening of the contractional sector.

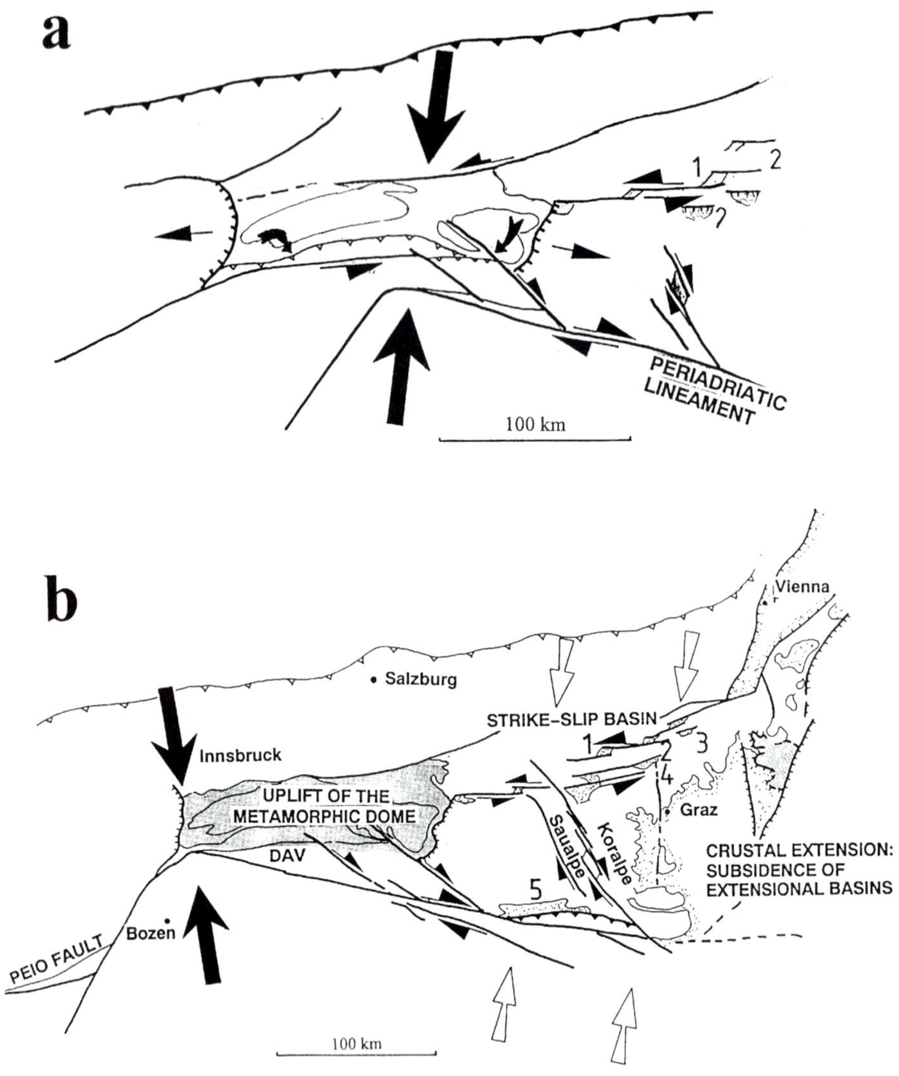

Fig. 4. Structures within the extruding wedge during (a) the stage of laterally unconfined margin and (b) during final general shortening (modified from Neubauer, 1994). Numbers locate sedimentary basins shown in Fig. 5. 1 - Trofaiach basin; 2 - Parschlug-Kapfenberg basin; 3 - Waldheimat basin; 4 - Leoben basin.

Using the above outlined model the ZMS is represented by the Tauern window section where ESE- and ENE-trending dome structures are related to possible blind thrusting at depth (Kurz and Neubauer, 1996). Furthermore, intense folding of the Austroalpine units south of the Tauern window may also be related to contraction. The thickening of crust is also shown by the deep level of the Moho at ca. 55 kilometres in contrast to more eastern sections where the Moho rises to ca. 25 kilometres beneath the Pannonian Basin (Aric et al., 1987). The ZMS within the Tauern window section is formed by oblique NNE-SSW contraction caused by indentation of the Southalpine block along central sectors of the Eastern Alps. This led to shortening by folding in high structural levels, transtensive oversteps along sinistral, orogen-parallel strike-slip faults, unroofing of lower plate crust by orogen-parallel (c. E-W), ductile, low-angle normal faults within the Tauern metamorphic core complex at the structural tip of the Southalpine indenter (Genser and Neubauer, 1989). Earlier NE-SW contraction is documented by activation of sinistral, E- to ENE-trending strike-slip faults, and by formation of NE-, later NNE-trending extensional veins within the Tauern window section (Kurz et al., 1994). Furthermore, blind floor thrusts at the base of fault-propagation folds may have significantly contributed to shortening and updoming of the Tauern metamorphic core complex (Kurz and Neubauer, 1996).

Strike-slip faults along the northern margins of the extruding wedge are associated with a number of different types of basins that were virtually formed and were infilled with clastic sediments during the extrusion process (Fig. 5). These basins include the following types: 1) Pull-apart basins, as the Trofaiach and Parschlug-Kapfenberg Basins (Fig. 5a; Neubauer, 1988; Ratschbacher et al., 1991b); 2) extensional basins in more eastern sectors that formed as half-grabens along ENE-trending, S-directed normal faults, as the Waldheimat and the Leoben basin (Fig. 5b, c). All these basins initiated and formed during Early Miocene extrusion. Most of them were later deformed by N-S, post-Middle Miocene shortening that resulted in folding and reverse faulting (e.g., Fig. 5b, c). These basins contrast with a basin along the Periadriatic fault, the Klagenfurt Basin, that initiated and formed during later, Sarmatian to Pleistocene, stages (Fig. 5d) and displays general contraction within more eastern sectors of the extrusion wedge (Polinski and Eisbacher, 1992). Contraction in the sector east of the Tauern window is also documented by large-scale, E-trending folds (Fig. 6) that also involve Neogene basins.

Fold lengths on the scale of 20 kilometres are interpreted to result from the young thermomechanical age of that unit (in comparison with models provided by Cloetingh and Burov, 1996). The second stage represents an ongoing tectonic process as documented by distribution of earthquakes and their fault plane solutions (Gutdeutsch et al., 1987), and in situ-stress measurements. These document ongoing general NNW-SSE contraction in the Alpine-Pannonian system including the former extrusional wedge (Becker, 1993; Horvath and Cloetingh, 1996; Peresson and Decker, 1997).

A marked contribution to late stages of extrusion was made by gravity that resulted from contraction at the tip of the indenting Southalpine block. This

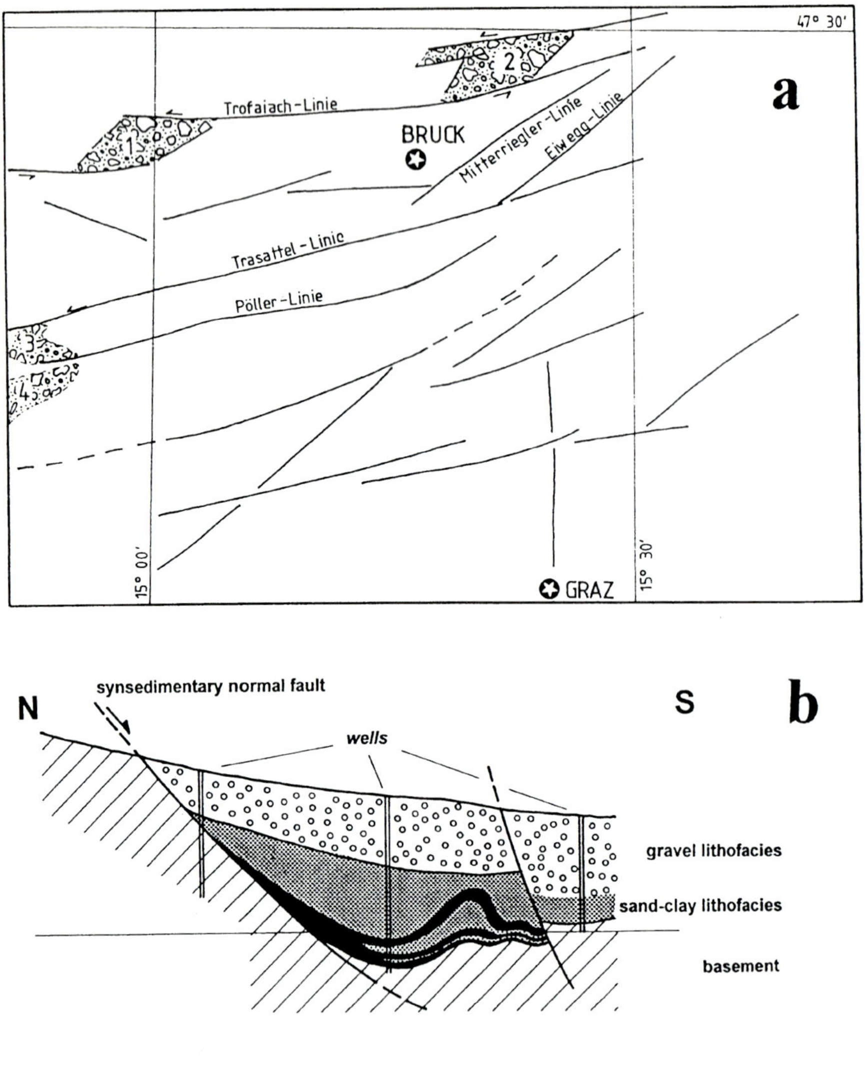

Fig. 5. Cross-section of different types of sedimentary basins during progressive evolution of an extruding wedge: (a) - The Trofaiach (1) and Parschlug-Kapfenberg (2) Basins that formed as pull-apart basins at oversteps on sinistral, E-trending strike-slip faults (modified from Neubauer, 1988). (b) - The Waldheimat Basin formed as a half graben during stage 1 (N-S extension with a high-angle, S-dipping normal fault along the northern margin. The basin was later contracted during stage 2, perpendicular to the displacement direction of the extruding wedge. Modified from Weber and Weiss (1983). For location of the basin, see Fig. 4b (no. 3).

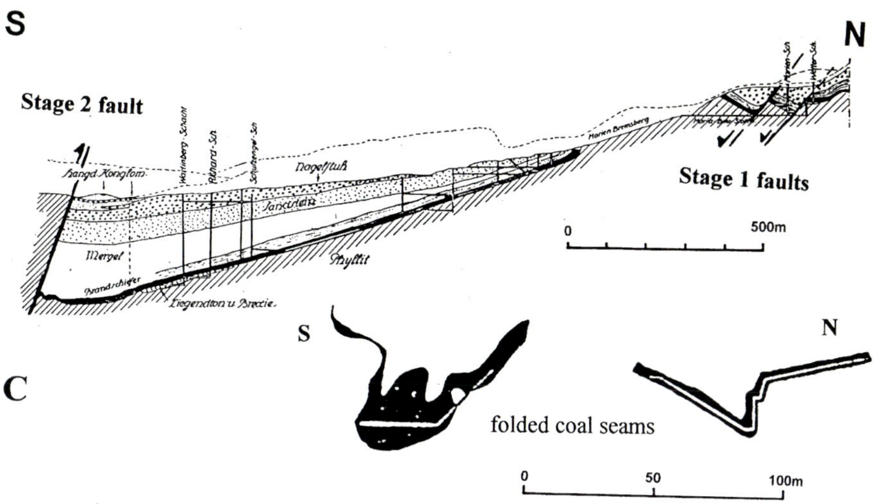

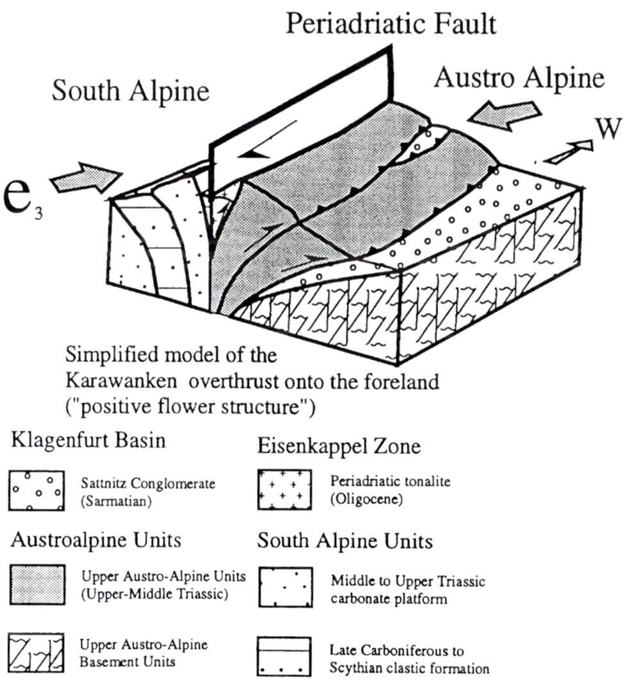

d

Fig. 5 (contimued). Cross-section of different types of sedimentary basins during progressive evolution of an extruding wedge: c - The Leoben Basin that represents a half graben too. The basin was formed as halfgraben by stage 1 normal faults, and inverted by later, stage 2 reverse faults and folding. Modified from Petrascheck, 1924. d - The Klagenfurt Basin, an intra-orogenic foreland basin that formed during stage 2 contraction along the southern wrench corridor (modified from Nemes et al., 1997).

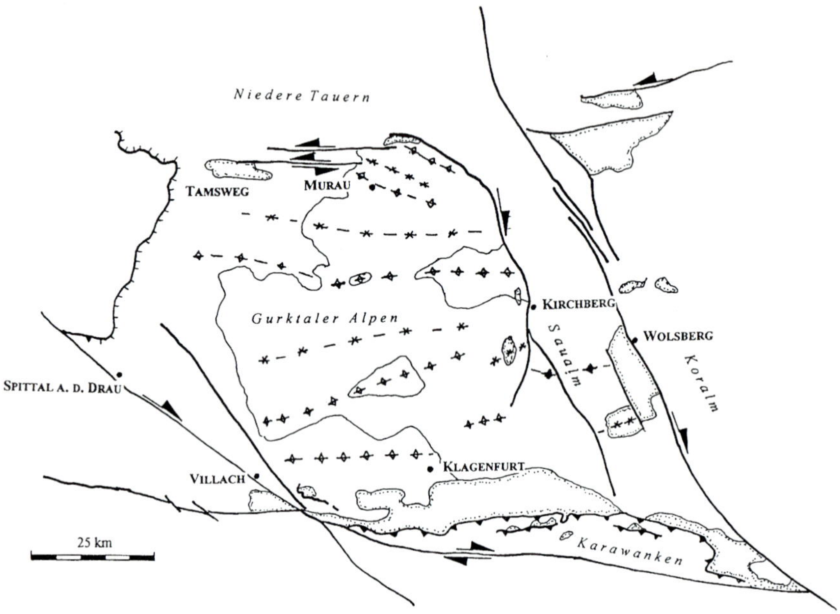

Fig. 6. Lithospheric-scale, large-scale, low amplitude folding within the extruding wedge (east of the Tauern window). Note that Neogene sedimentary basins (stippled areas) are involved. Folding postdates, therefore, Miocene basin formation.

induced surface uplift at the western tip of the extrusional wedge and, therefore, gravitational instability of the wedge. The uplift, and block tilting of upper, brittle sectors of the extrusional wedge may have significantly contributed to the extrusion. Ductile low-angle normal faults led to the detachment of upper brittle sectors of the crust from lower crustal rocks due to thermally and lithologically controlled strain concentration within the orogenic wedge. Middle and lower crustal rocks were ductilely deformed and flowed more rapidly to the east, out of the contractional sector, than did upper, brittle portions. This effect led to the juxtaposition along low angle normal faults of brittle, upper crust with highly ductilely deformed middle sectors of the crust. Associated crustal thinning also resulted in a thermal perturbation of upper plate rocks (Sachsenhofer, 1992). The master low-angle normal fault is the E-dipping, and ESE-displacing ductile normal fault along the eastern margins of the Tauern window (Fig. 4a, b) that also forms a mechanical separation of the brittle upper sectors from the ductile lower sectors of the extrusional wedge. Similar ductile low angle normal faults are exposed along upper margins of the Penninic Rechnitz windows group within the Pannonian Basin (Fig. 4b). This ductile low angle normal fault resulted in top-to-the-west sense of ductile shear along upper margins of the exhuming Rechnitz metamorphic core complexes within the extruding wedge. The relationships between ductile and brittle segments within eastern sectors of the extrusional wedge imply, therefore, relatively seen, opposite ductile rock flow in lower and middle sectors of the crust. The interior of both the Tauern and the Rechnitz windows is largely controlled by ductile pure shear and east-west stretching of the lithosphere (Kurz et al., 1994; Ratschbacher et al., 1992). E-W extension is also documented by formation of extensional basins within the Pannonian basin system (Royden and Horvath, 1988; Bergerat et al., 1989; Tari et al., 1992).

Acknowledgements: We appreciate discussion and long influence by Georg Mandl to the combined Graz/Salzburg structural working groups. Work has been supported by grants 8652 and 9918 of the Austrian Science Foundation to FN.

References

Aric, K., Gutdeutsch, R., Klinger, G. and Lenhardt, W. (1987) Seismological studies in the Eastern Alps. - In: Flügel, H.W. and Faupl, P., eds., Geodynamics of the Eastern Alps. pp. 325-333, Deuticke, Vienna.

Balla, Z. (1985) The Carpathian loop and the Pannonian basin: A kinematic analysis. - Geophys. Trans., 30, 313-353.

Becker, A. (1993) Contemporary state of stress and neotectonic deformation in the Carpathian-Pannonian region. - Terra Nova, 5, 375-388.

Bergerat, F. (1989) From pull-apart to the rifting process: the formation of the Pannonian basin. - Tectonophysics, 157, 271-280.

Cloetingh, S. and Burov, E. (1996) Thermomechanical structure of European continental lithosphere: constraints from rheological profiles and EET estimates. Geophys. J. Int.,124 695-723.

Fodor, L. (1995) From transpression to transtension: Oligocene-Miocene structural evolution of the Vienna basin and the Eastern Alps/Western Carpathian junction. - Tectonophysics, 242, 151-182.

Genser, J. and Neubauer, F. (1989) Low angle normal faults at the eastern margin of the Tauern window (Eastern Alps). - Mitt. Österr. Geol. Ges., 81, 233-243.

Gutdeutsch, R. and Aric, K. (1987) Tectonic block models based on the seismicity in the East Alpine-Carpathian and Pannonian area. - In: Flügel, H.W. and Faupl, P., Geodynamics of the Eastern Alps, pp. 309-324; Deuticke, Vienna.

Horvath, F. and Cloetingh, S. (1996). Stress-induced late-stage subsidence anomalies in the Pannonian basin. Tectonophysics, 266, 287-300.

Kurz, W., Neubauer, F., Genser, J. and Horner, H., (1994): Sequence of Tertiary brittle deformations in the Eastern Tauern window (Eastern Alps). - Mitt. Österr. Geol. Ges., 86, 153-164.

Kurz, W. and Neubauer, F. (1996) Deformation partitioning during updoming of the Sonnblick area in the Tauern window (Eastern Alps, Austria). J. Struct. Geol., 18, 1327-1343.

Nemes, F., Pavlik, W. and Moser, M. (1995) Geologie und Tektonik im Salza-Tal (Stmk.) - Kinematik und Paläospannungen entlang des Ennstal-Mariazell Blattverschiebungssystems in den Nördlichen Kalkalpen. - Jb. Geol. Bundesanst., 138, 349-367.

Nemes, F., Neubauer, F., Cloetingh, S. and Genser, J. (1997) The Klagenfurt basin in the Eastern Alps: an intra-orogenic decoupled flexural basin? - Tectonophysics, 282 , 189-203.

Neubauer, F., (1988) Bau und Entwicklungsgeschichte des Rennfeld-Mugel- und des Gleinalmkristallins. Abh. Geol. Bundesanst., 42: 1-137.

Neubauer, F. (1994) Kontinentkollision in den Ostalpen. -Geowissenschaften, 12: 136-140.

Neubauer, F. and Genser, J. (1990) Architektur und Kinematik der östlichen Zentralalpen - eine Übersicht. - Mitt. naturwiss. Ver. Steiermark, 120: 203-219.

Peresson, H. and Decker, K. (1997) Far-field effects of late Miocene subduction in the Eastern Carpathians: E-W compression and inversion of structures in the Alpine-Carpathian region. - Tectonics, 16 , 38-56.

Petrascheck, W. (1924) Kohlengeologie der österreichischen Teilstaaten. VI. Braunkohlenlager der österreichischen Alpen. - Berg Hüttenmänn. Mh., 72, 5-48, 62-101.

Polinski, R. K. and Eisbacher, H. G. (1992) Deformation partitioning during polyphase oblique convergence in the Karawanken Mountains, southeastern Alps. - J. Struct. Geol., 14, 1203-1213.

Ratschbacher, L., Behrmann, J. and Pahr, A. (1990) Penninic windows at the eastern end of the Alps and their relation to the intra-Carpathian basins. - Tectonophysics, 172, 91-105.

Ratschbacher, L., Merle, O., Davy, Ph. and Cobbold, P. (1991a) Lateral extrusion in the Eastern Alps. Part 1: Boundary conditions and Experiments scaled for gravity. - Tectonics, 10, 245-256.

Ratschbacher, L., Frisch, W., Linzer, H.G. and Merle, O. (1991b) Lateral extrusion in the Eastern Alps. Part 2: Structural analysis. Tectonics, 10, 257-271.

Royden, L.H. and Horvath, F., eds. (1988): The Pannonian Basin. A Study in Basin Evolution. - Amer. Ass. Petrol. Geol. Mem., 45.

Sachsenhofer, R. F. (1992) Coalification and thermal histories of Tertiary basins in relation to late Alpidic evolution of the Eastern Alps. - Geol. Rdschau, 81, 291-308.

Sprenger, W. and Heinisch, H. (1988) Late Oligocene to Recent brittle transpressive deformation along the Periadriatic Lineament in the Lesach Valley (Eastern Alps): remote sensing and paleo-stress analysis. - Annales Tectonicae, VI/2, 134-149.

Tapponnier, P. and Molnar, P. (1976) Slip-line fields theory and large-scale continental tectonics. - Nature, 264, 319-324.

Tari, G., Horvath, F. and Rumpler, J. (1992) Styles of extension in the Pannonian Basin. - Tectonophysics, 208, 203-219.

Wang, X., Nemes, F. and Neubauer, F. (1996) Polyphase kinematics of strike-slip faulting along the Salzachtal-Ennstal fault, Eastern Alps. - In: Amann, G., Handler, R. Kurz, W. and Steyrer, H.P., eds., 6. Symposium Tektonik - Strukturgeologie - Kristallingeologie, Salzburg 10.-15. April 1996, Erweiterte Kurzfassungen. Facultas-Universiätsverlag, Vienna, 466-469.

Weber, L. and Weiss, A. (1983) Bergbaugeschichte und Geologie der österreichischen Braunkohlenvorkommen. Archiv Lagerstättenforsch. Ostalpen, 4: 1-317.

In-situ stress measurements in model experiments of tectonic faulting

Dick A. Nieuwland[1,2], Janos L. Urai[1,3], and Maaike Knoop[1,4]

[1] Shell International Exploration and Production BV, PO Box 60, 2288BV Rijswijk The Netherlands

[2] now at: Faculty of Earth Sciences Vrije Universiteit Amsterdam, De Boelelaan 1085,1081 HV Amsterdam, The Netherlands

[3] now at: Geologie-Endogene Dynamik, RWTH Aachen, Lochnerstrasse 4-20, D-52056 Aachen, Germany.

[4] present adress: Kruiser 8, 3904 ZR Veenendaal, The Netherlands

Abstract. Measurement of in-situ stresses during an experiment has been long on the wish-list of sandbox modellers. This paper reports the first results of such measurements, using miniature solid state pressure transducers. These allow a quantification of the development of stress fields during model experiments of thrust tectonics, and form a basis for comparing sandbox experiments with theoretical models.
First results show that within each experiment the pattern of variation in horizontal stress in space and time is consistent. The stress magnitudes are reproducible and in agreement with the expected values. Low amplitude stress cycling after the initiation of a fault is an interesting observation, interpreted to represent stick-slip process in the fault zone. The measured variations in horizontal stress can also be correlated to density variations seen in X-ray Tomograph (CT) scans. This provides additional means to visualize aspects of a stress field development in 3D and in time.

1 Introduction

The use of sandbox experiments (Horsfield, 1977; Mandl, 1988; Richard et al., 1995) to model the tectonic evolution of the Earth's crust depends critically on the correspondence of deformation structures in the model to those in the model's prototype.

Although the general fault pattern and the main structure of the stress field are thought to be reasonably understood, other structures in the model may not correspond to those in the prototype (Horsfield, 1977, Leroy and Triantafylidis, this volume). Assessment of these uncertainties can be done by comparing measured displacements and forces in the model with results of theoretical models. For this purpose the measurement of in-situ stresses during an experiment has long been on the wish list of sandbox modellers.

In geotechnical practice in situ stresses in sand and clay are measured with pressuremeters (Kulhawy, 1982; Dunnicliff and Green, 1987). Attempts to apply these techniques to sandbox experiments have not been succesful to date due to the large dimensions of geotechnical pressuremeters and the very low stress magnitudes in the sandbox. In this paper we report the application of miniature solid state pressuremeters to measure the stresses in sandbox experiments of thrust tectonics. The measurements are compared with values expected from simple mechanical considerations, and with simultaneous density variations in time-lapse X-ray Tomograph scans of the deforming models.

2 Calibration

The basic procedures used in this study follow those outlined in Weiler and Kulhawy (1982) and Dunnicliff and Green (1987) for large pressuremeters in geotechnical practice. Besides calibration in a fluid, the use of a pressure cell in sandbox experiments requires calibration of the sensor against a known stress field in sand, selection of a sensor with suitable size and stiffness (Harris and Seager 1973), and careful design of the emplacement procedure to obtain maximum reproducibility.

Experimental procedures

We used miniature solid state pressure transducers (Kyowa PS-2KA), calibrated in air to be linear and accurate to within 1% over the pressure range of interest (0 - 5000 Pa). The transducers are thin circular plates with a thickness of 0.6 mm and a diameter of 6 mm. When loaded in the sandbox, the ratio of diameter to deflection is about 10^5. A fundamental, as yet untested assumption of this study is that in sand the sensors register the component of the local stress vector which acts normal to the sensor, independent of the value of the shear stress. To describe the orientation of the sensor, we specify the normal to the sensor's plane, i.e. a horizontal sensor is embedded in the sand with the sensor plane vertical, to register the horizontal stress. A small cylindrical test cell was designed for the calibration (Fig.1) In this cell, the horizontal stress in the sandpack was controlled with a thin membrane driven by compressed air. This setup enabled the creation of an accurately controlled homogeneous stress field in the central part of the sand pack. The homogeneity of the stress field in the centre of the cell was confirmed by an elasto-plastic finite-element analysis, using the program PLAXIS (Fig. 1b).

In a calibration test the chamber is first filled half by sprinkling sand from 1 m height. Then the horizontal sensor is inserted, and the cell is filled up with sand completely, taking care not to move the sensor in the process. After allowing temperature to stabilise, the horizontal stress (i.e. the pressure behind the membrane) is slowly varied and plotted against the corresponding output of the pressure transducer in the sand. Some fifty calibration experiments in air and in sand were carried out, systematically varying details of the procedure (e.g. exact position of the sensor, filling procedure, type of membrane). In this study we only carried out calibrations with the sensor's normal at right angles

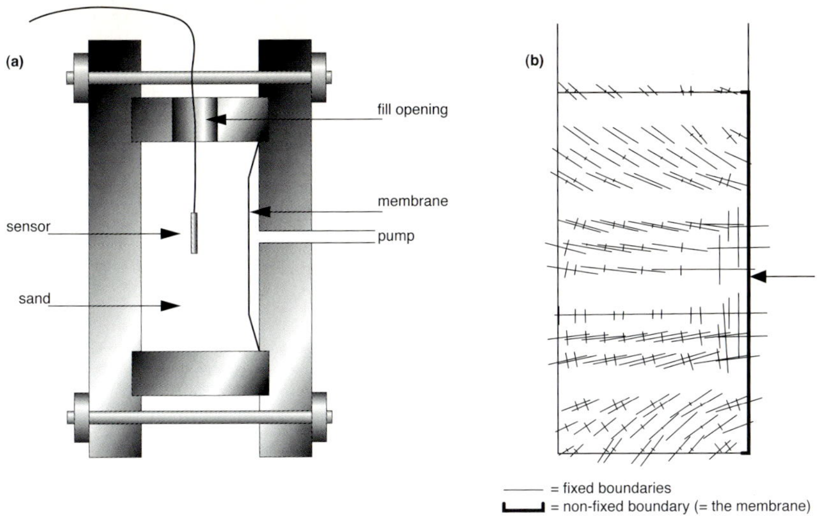

Fig. 1. a) Schematic drawing of the cylindrical calibration cell. b) Principle stresses in the sand pack caused by loading through the membrane, as calculated using the finite element program Plaxis (stresses around the sensor not modelled). In the centre of the cell the principle stress is nearly horizontal and its magnitude is very close to the pressure of the membrane.

to the membrane, i.e. paralell to a principle stress. Further work is needed to test the assumption of the sensor's insensitivity to shear stresses.

3 Calibration results

Typical calibration curves are shown in Figure 2. The hysteresis loops seen are of the type found by other workers attempting to measure stress in granular materials. Keeping the preparation procedures and temperature carefully constant, the results are quite reproducible by cyclic loading within one experiment, but the average slope of the loop varies from experiment to experiment, and depends on the details of the calibration procedure. Although no clear systematic effect of these parameters could be demonstrated, for the final procedure the average slope of the loop is described by $P_{sensor} = 0.9 \pm 0.1 \times P_{membrane}$. Thus in general the sensor output is slightly lower than the horizontal stress in the sand. As shown by the shape of the hysteresis loops, at the beginning of the loading the sensor underregisters more.

The development of horizontal stress (σ_h) with increasing overburden thickness was investigated to a sandpack thickness of some 40 cm above a horizontal sensor placed in a 40 cm wide container with rigid vertical walls (uniaxial compaction). Vertical stress (σ_v) was estimated from sandpack height and density, and the horizontal stress was measured by the sensor. The horizontal stress was found to increase linearly with overburden stress and the value of K_o ($=\sigma_h/\sigma_v$)

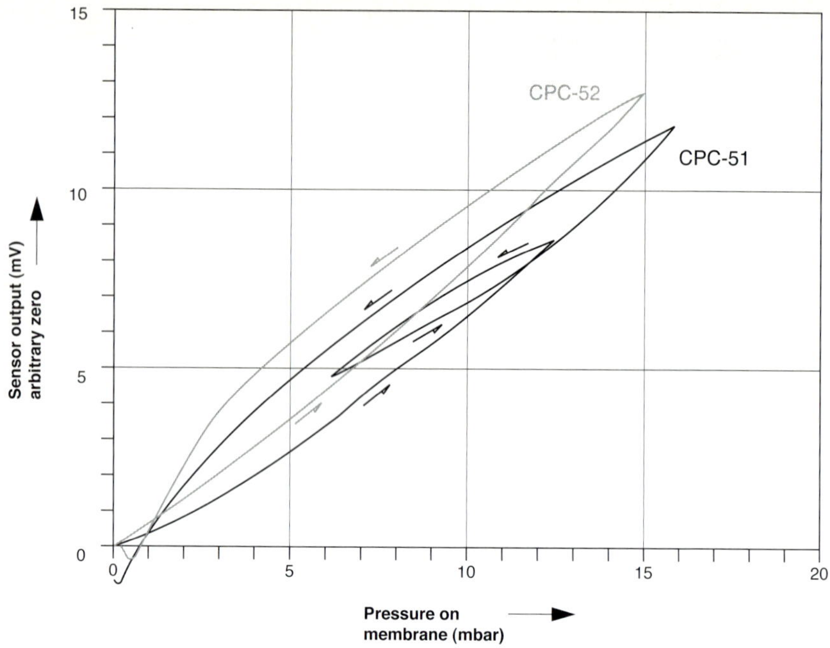

Fig. 2. Calibration curves (1 mbar = 100 Pa.

was found to be around 0.5, in reasonable agreement with values reported in the literature for sand (Mitchell, 1976).

One surprising observation was, that our technique did not seem to work for vertical sensors: in this orientation the sensor output was far below the calculated overburden stress. This needs further study, as Harris and Seeger (1973) did obtain satisfactory response for vertical sensors. Therefore at this point our setup is only capable of measuring the horizontal stress in a sandpack.

4 Measurements in thrust experiments

The first series of measurements were carried out in model experiments of thrust tectonics (the strong increase of the horizontal stress when the sandpack is shortened gives the best signal to noise ratio). The sandpack (3 cm thick, without lubrication along the base) was shortened horizontally by a moving backstop, in an approximately plane strain deformation. One horizontal stress sensor with the sensor normal paralell to the shortening direction was built into the model. The measured stress therefore corresponds to the horizontal component of the in-situ stress tensor (σ_{xx}, when the coordinate system is chosen with the x-axis paralell to the displacement vector of the backstop, and the y-axis is vertical. To determine the four components of the 2D stress tensor in this coordinate system, measurements on three differently oriented sensors are required, see Appendix).

The results of the first experiment (test T2) are shown in Fig. 3. The development of the horizontal stress (σ_{xx}) with time is characterised by a rapid increase until a peak stress, followed by sharp drop, leading to a phase with oscillation of σ_{xx} around a constant mean value. This cycle is repeated twice during progressive shortening.

We interpret the steep rise in σ_{xx}, to represent the build-up of the horizontal stress, leading to the formation of the first thrust fault. The full magnitude of the stress rise is not yet monitored by the sensor because of the high basal friction and the distance between the sensor and the fault (Hafner, 1951; Mandl, 1988 p. 183). The sharp drop in σ_{xx} followed by oscillations represents the overthrust phase, whereby the newly formed thrust fault is observed to break through at the surface shortly after the peak stress is reached. When the thrust sheet begins to overrun the footwall, the small but regular variations in σ_{xx} are interpreted to be a consequence of stick-slip motion on the fault. In later experiments where several sensors were placed into the model along strike, the stress cycles showed excellent correlation between sensors (we carefully eliminated stress oscillations in the loading system, for example by friction of the backstop). The second much larger increase in σ_{xx} registers the formation of the second overthrust fault, (this time much closer to the sensor). The further increase in σ_{xx} (leading to the third thrust past the sensor) is interpreted to reflect the increased thickness of the sand layer above the sensor.

In the second, similar experiment (T3, Fig. 4), this sequence is altered slightly due to the formation of a backthrust before the formation of the third thrust. This results in a double peak.

Fig. 5 shows the stress state around the sensor at different stages of this experiment, in a Mohr diagram. Initial stresses (with $\sigma_{xx} < \sigma_{yy}$) were estimated from overburden thickness and K_0 as discussed above. As the second thrust was formed at a distance from the sensor, the measured stress is lower than the values near the thrust itself. The third thrust is initiated close to the sensor, and at this point the overburden stress has also increased slightly. The measured values were compared with the failure envelopes for sprinkled sand reported by Krantz (1991). A value for cohesion of about 500 Pa seems reasonable for the sandpack used in our experiments, however the friction angle of 45 degrees is neither consistent with our stress measurements nor with the dip of the thrusts of around 30 degrees, (corresponding with a friction angle of 30 degrees for horizontal σ_1). Therefore we believe that friction angle in our sandpacks is significantly lower than the values suggested by Krantz (1991). In Fig. 5 we plotted a failure envelope with $C = 500$ Pa and friction angle of 35 degrees for comparison.

In a third experiment, we used three sensors placed at increasing distance from the backstop (Fig. 6). The output of the three sensors follow a pattern similar to the previous ones, but the increase in distance from the thrust front is expressed in a decrease in magnitude of the measured horizontal stress σ_{xx}. After the formation of the second thrust at around 500 seconds, the response of the three sensors becomes similar. The formation of the third thrust is marked by a sharp increase followed by a sharp drop in the output of the first sensor:

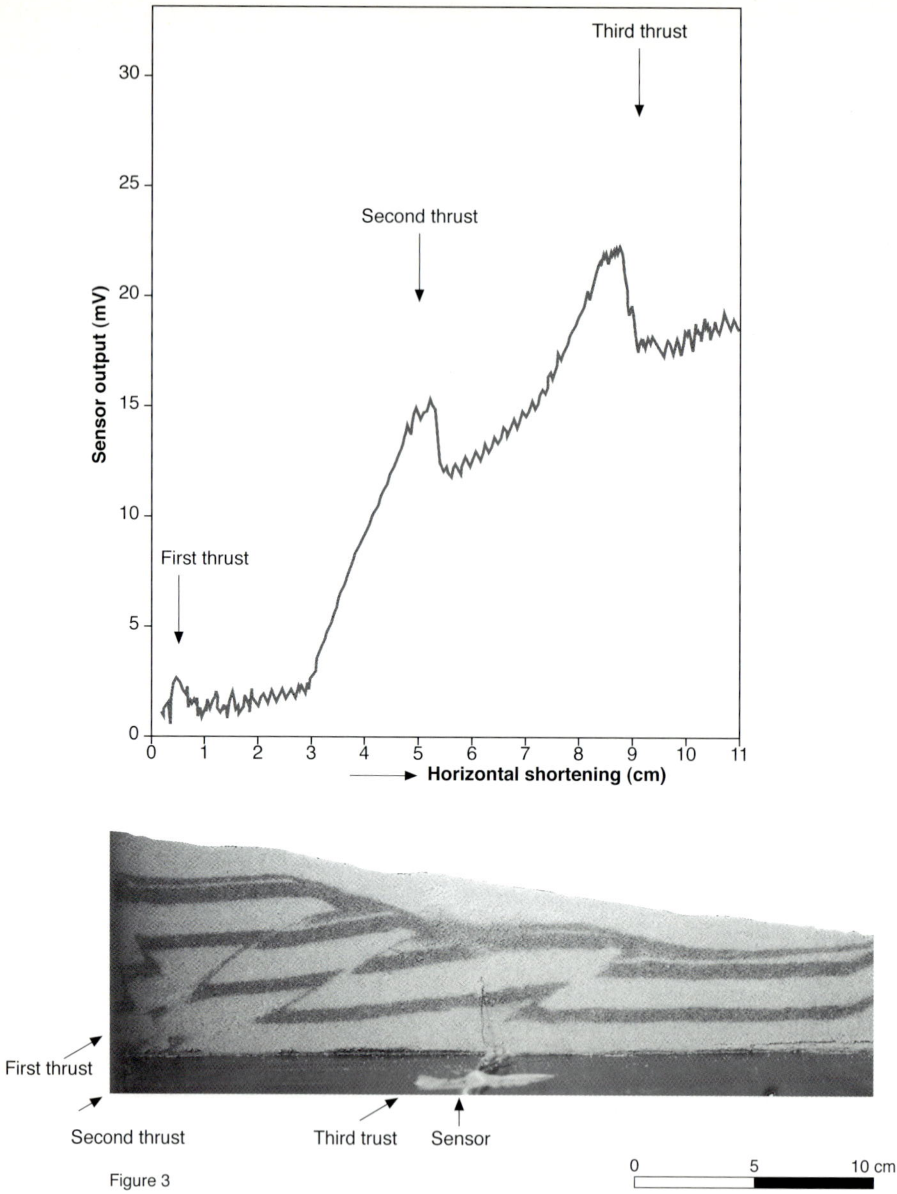

Fig. 3. Horizontal stress against time in a sandbox experiment of thrust tectonics (T2), as indicated by the sensor output. Insert shows a profile of the sectioned model after the experiment. Position of the sensor is indicated. Sensor output is set to zero after the two hour stabilisation phase, the magnitude of the stress is thus measured relative to the in-situ stress before the onset of the deformation. Note that conversion of the sensor output (using the calibration data) to obtain the true horizontal stress was not done.

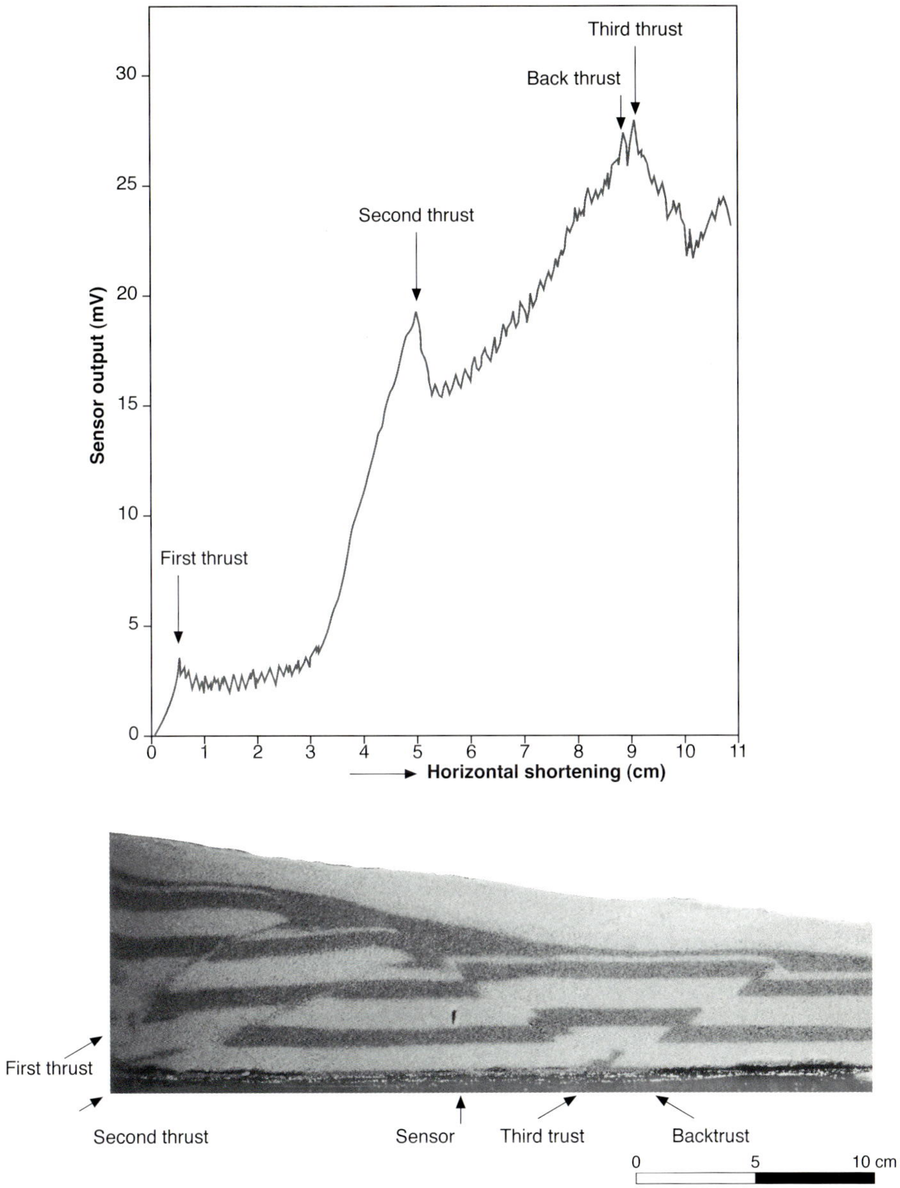

Fig. 4. Horizontal stress against time in a sandbox experiment of thrust tectonics (T3), insert shows profile of the sectioned model after the experiment. Note that conversion of the sensor output (using the calibration data) to obtain the true horizontal stress was not done.

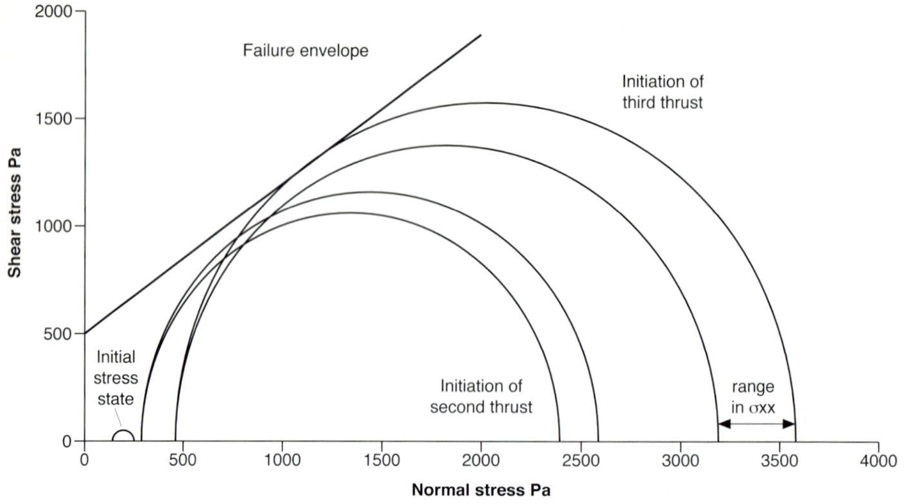

Fig. 5. Mohr diagram of the stresses in the experiment shown in Fig. 4

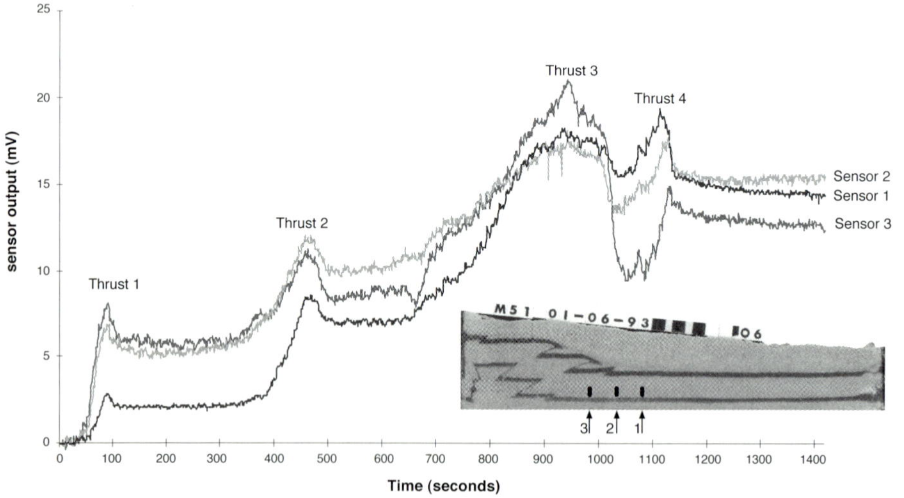

Fig. 6. Horizontal stress against time in a sandbox experiment of thrust tectonics (M51), insert shows profile of the sectioned model after the experiment. This experiment was run with three sensors in place, showing how the stress field varies in space and time.

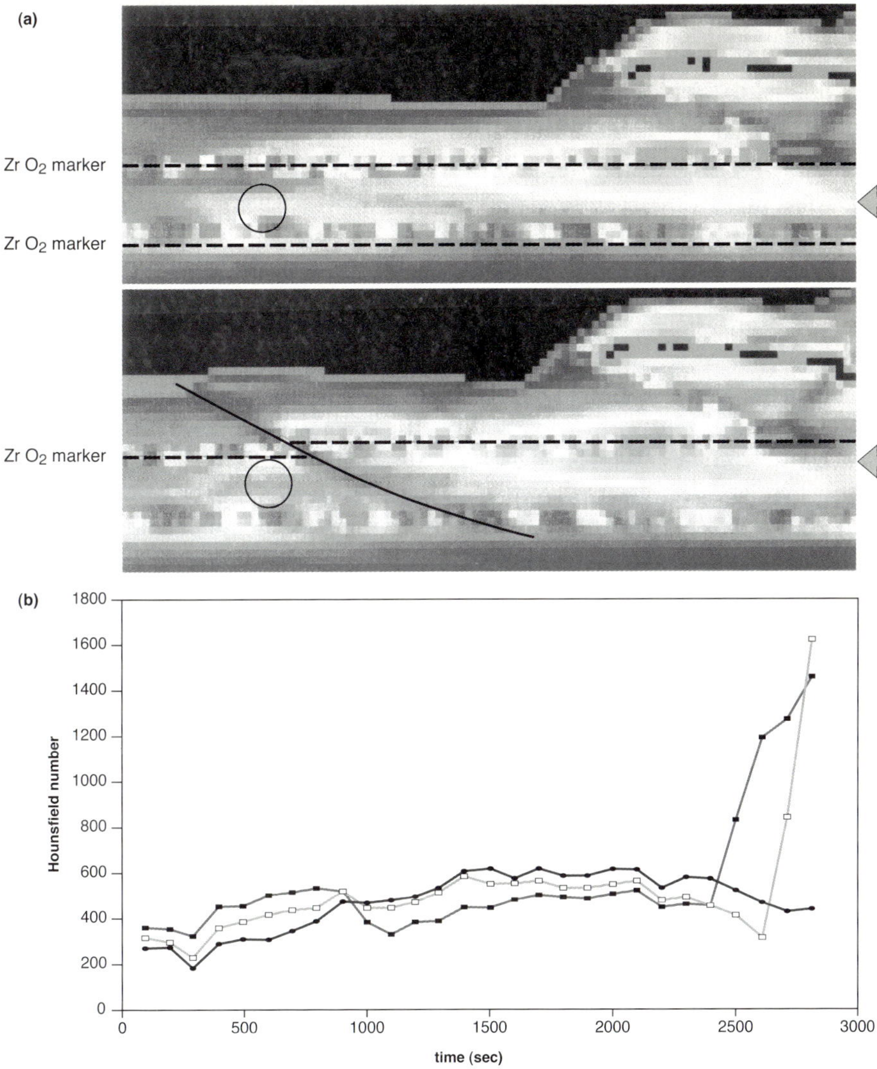

Fig. 7. (a) Density fluctuations in front of a thrust fault as observed with X-ray tomography. The regions of interest on the cross sections indicate the positions of the corresponding density increase (light colored) and subsequent decrease (dark colored) in the same position as where the stress sensor was placed. The dark blue coloring along the thrust fault trajectory represents a density decrease due to dilatation. Two layers of zirconium-oxide were added as markers. (b) Radiological density (represented by Hounsfield number) plotted against time for three voxels from the marked area. A density decrease is apparent, similar to the stress drop as measured with the stress sensors and as indicated on the images.

after sectioning the model it was shown that this sensor lies exactly in the fault plane and the deviation in response was caused by the large nonuniform strains (and corresponding bending moments) around the sensor.

5 Computer Tomograph scans

A number of experiments similar to the ones described above were done in a Computer Tomograph (CT) scanner, yielding a series of 3D images of both the geometrical development of the structure with time and the model's density distribution. Density fluctuations over time at selected locations in the model, were observed to show similar development as the stresses measured by the stress sensors (Fig. 7). The increase in σ_{xx} prior to thrust break-through and the sharp drop in σ_{xx} after the appearance of the thrust fault at the surface, correspond to a density increase followed by a density drop in the region directly in front of the developing thrust fault (Fig. 7). Density of selected volumes of the model, extracted at 200 sec intervals, illustrates this behaviour.

Local changes in density of the sandpack are directly related to the development of the volumetric strain field. This in turn is related to the stress history, albeit in a complicated and nonlinear way with initial compaction due to the large increase in mean stress, followed by shear-induced dilatancy. Although this relationship is complicated and much further work is needed, 3D density surveys, acquired at different stages of the development of a model, have the potential to provide additional indication of the development of the in-situ stress field.

6 Conclusions

- In-situ stress measurements in model experiments of tectonic faulting can be carried out succesfully. The results are in agreement with expectations and are reproducible.
- The stress cycling during the development of stress fields in 3D and in time are more complicated than expected from simple Mohr-Coulomb plasticity, and form the basis for comparison with theoretical models.

Appendix

Determination of the components of the stress tensor (in 2D) from stress cell measurements in the sandbox.

Assumption: Shear traction on the sensor has no effect on the measurements which correspond to the normal stress on the sensor. Let σ_1 and σ_2 be the principle stresses of the stress tensor σ in the point under consideration, and θ the angle between a plane's normal vector and σ_1. The normal component N of the traction vector on this plane is then given by

$$N = \sigma_1 \cos^2 \theta + \sigma_2 \sin^2 \theta \quad \text{or} \quad 2N = (\sigma_1 + \sigma_2) + (\sigma_1 - \sigma_2) \cos 2\theta \tag{1}$$

If N is measured on three different planes (giving $N^{(1)}$, $N^{(2)}$, and $N^{(3)}$ respectively) the 2D stress tensor can be reconstructed. Let α and β be the angles between the normals to the first and second and first and third sensor, respectively. We have then the following equations:

$$2N^{(1)} = (\sigma_1 + \sigma_2) + (\sigma_1 - \sigma_2)\cos 2\theta \tag{2}$$
$$2N^{(2)} = (\sigma_1 + \sigma_2) + (\sigma_1 - \sigma_2)\cos(2\theta + 2\alpha) \tag{3}$$
$$2N^{(3)} = (\sigma_1 + \sigma_2) + (\sigma_1 - \sigma_2)\cos(2\theta + 2\beta) \tag{4}$$

let

$$\frac{N^{(1)} - N^{(2)}}{N^{(1)} - N^{(3)}} = \frac{\cos 2\theta - \cos(2\theta + 2\alpha)}{\cos 2\theta - \cos(2\theta + 2\beta)} = C \tag{5}$$

then

$$\tan 2\theta = \frac{1 - C\cos 2\beta - \cos 2\alpha}{C\sin 2\beta - \sin 2\alpha} \tag{6}$$

and

$$(\sigma_1 + \sigma_2) = \frac{2N^{(2)}\cos 2\theta - 2N^{(1)}\cos(2\theta + 2\alpha)}{\cos 2\theta - \cos(2\theta + 2\alpha)} \tag{7}$$

and

$$(\sigma_1 - \sigma_2) = \frac{2N^{(1)} - (\sigma_1 + \sigma_2)}{\cos 2\theta} \tag{8}$$

References

Dunnicliff, J., and G.E. Green.(1981). Geotechnical instrumentation for monitoring field performance. Wiley Interscience.

Harris, G.W. and Seager, J.S. (1973) Pressure measurement in sand model experiments and the use of pressure sensitive transistors. International Journal of Rock mechanics and Mining Science and Geomechanical Abstracts 10:613-622.

Hafner, W. (1951) Stress distribution and faulting. Bulletin of the Geological Society of America 62:373-398.

Horsfield, W.T. (1977) An experimental approach to basement controlled faulting. Geologie en Mijnbouw 56:363-370.

Jewell, R.J. Wroth, C.P. Windle, D. (1980) Laboratory studies of the Pressuremeter tests in sand. Geotechnique 4:507-531.

Jones, M.E. and Addis M.A., (1986) The application of stress path and critical state analysis to sediment deformation. Journal of Struct. Geology, V.8, nr.5. p. 575-580.

Krantz, R.W. (1991) Measurements of friction coefficients and cohesion for faulting and fault reactivation in laboratory models using sand and sand mixtures. Tectonophysics 188:203-207.

Leroy, Y.M. and Triantafillidis, N. (1997) Stability of a frictional, cohesive layer on a viscous substratum: validity of assymptotic solution and influence of material properties. Journal of Geophysical Research, 101 (B8) 17795-17811.

Mandl, G. (1988) Mechanics of tectonic faulting. Vol. 1 in: Developments in structural geology, H. J. Zwart (editor). Elsevier, 407 pp.

Mitchell, J.K. (1976) Fundamentals of Soil Behavior. Edited by T. W. Lambe, Series in Soil Engineering. New York: John Wiley & Sons, Inc.

Nieuwland, D.A. Walters, J.V. (1993) Geomechanics of the South Furious Field. An integrated approach towards solving complex structural geological problems, including analogue and finite element modelling. Tectonophysics 226:143-166.

Richard, P.D. Naylor M.A. Koopman A. (1995) Experimental models of strike-slip tectonics. Petroleum Geoscience 1:71-80.

Weiler, W.A., and F.H. Kulhawy. (1982) Factors affecting stress cell measurements in soil. Journal of Geotechnical Engineering Division, Proc. ASCE. 108 (GT12):1529-1548.

Development of joint sets in the vicinity of faults

Jean-Pierre Petit[1], Vincent Auzias[2], Keith Rawnsley[3], and Thierry Rives[3]

[1] Laboratoire de Géophysique et Tectonique, Université Montpellier II
[2] Total, Paris la Défense, France
[3] Elf Aquitaine Production, Pau, France

Abstract. Unlike regional joint patterns, fault-related joints can often be related to stress perturbations. Their patterns, which have seldom been described, are sometimes complex and difficult to decipher. The examples presented here are mainly joints associated with meter to decameter scale pre-existing faults in limestone. The joints mostly tend to be parallel or perpendicular to the faults with the occurrence of sets of diverging joints localized along the faults. The patterns presented are interpreted in the light of photoelastic models which enable the prediction of stress deviations linked to subtle morphological details on faults. These models show that many of the observed features can be interpreted as stress deviations linked to alternation of closed and open segments along the faults. Joints appear as excellent markers of the deviation of the major horizontal principal paleostress trajectories. A mechanical discussion based on the paradox of the absence of concentration of joints in zones of tensile stress concentration leads to the idea that joint propagation is assisted by contraction within the layers.

1 Introduction

Although joints are the most ubiquitous and abundant brittle structures in the earth's upper crust, they have in many respects remained mechanical enigmas. After much debate it is now generally agreed that the majority of joints are mode I (opening mode) fractures (Pollard and Aydin 1988). Some descriptive aspects have not been fully explored. Most descriptions of joint sets have concentrated on regional systems characterized by quite constant directions reflecting the presence of single or superimposed systematically orientated stress fields. Such regional systems are thought to have formed during main compressional or extensional deformation phases. However, this systematic orientation may be absent from folded and faulted zones and be replaced by different joint orientations. These may correspond to the same generation as the regional joints which develop in particular stress fields, and also to specific sets which cannot be related in time and space to the regional joint sets. The occurrence of deviated joints in the vicinity of faults is poorly studied and understood and is not a general phenomenon, as shown for example by Auzias (1995) in the Old Red Sandstones on the north coast of Scotland. Since faults induce complex perturbed stress fields, joint patterns related to stress trajectory deviations (with respect to the far or regional stress field) can be very complex and thus difficult to decipher.

Large high quality outcrops are necessary to observe these deviations, and these are often difficult to study without a helicopter view.

After a brief presentation of some classical joint associations with faults we will mainly concentrate on the description and mechanical interpretation of a few examples of joint perturbations along strike-slip faults all located around the Bristol Channel and along the Yorkshire coast. This paper (which includes some data already published and some from petroleum company reports) only aims to give a general outline of the outcrops and associated problems; more detailed studies are being done.

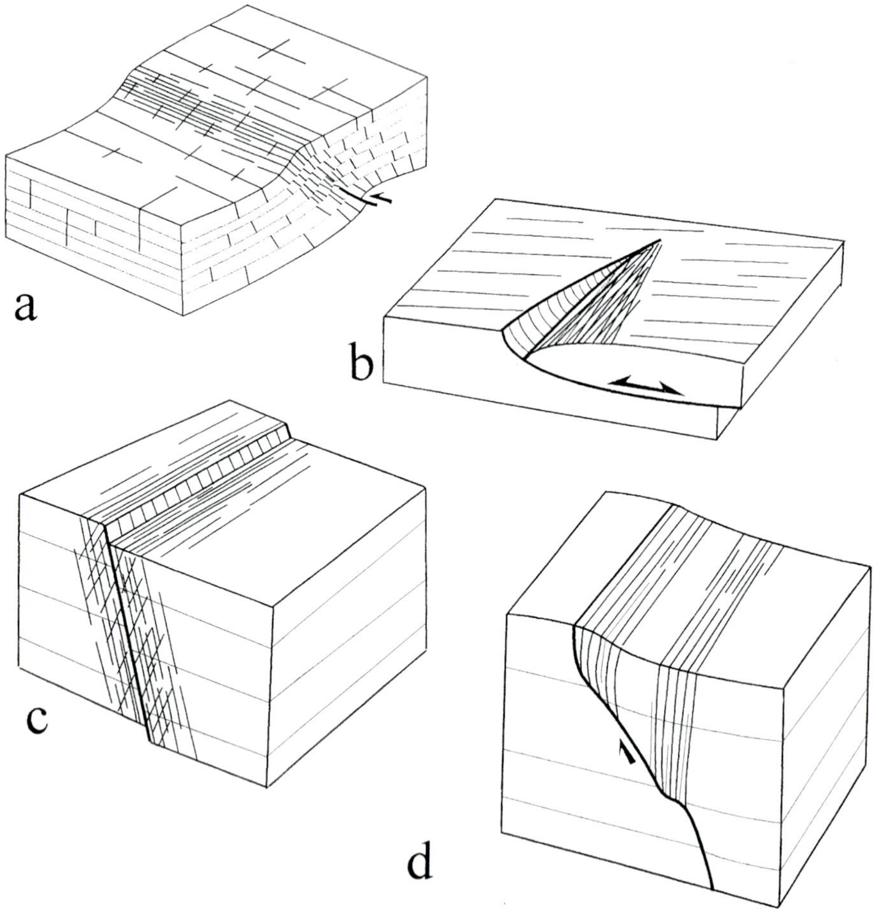

Fig. 1. Classically observed fault-joint relationships. a - relationship with cylindrical folding induced by an underlying fault; b- fan-shaped joints linked to a non cylindrical deformation on a rollover linked to an inverted listric fault; c - oblique and parallel joints along faults; d - swarms of joints localized at a fault bend and tip

2 Classical associations

Joint development can be related to faults either indirectly by the effect of folding associated with faults, or directly when joints are in direct contact with a faulted zone. The first type (Figure 1a) can be represented by orthogonal joint sets that develop within cylindrical flexures associated with underlying faults (Tricart et al. 1986). Another example comprising superposed joint sets of progressively varying orientations is described by Rives (1992) in the Liassic limestone of Lilstock (south of the Bristol Channel, England) where east-west trending joints could be linked to the inversion of a previously existing non-cylindrical roll-over anticline during the Pyrenean compression (Figure 1b).

An aspect of the second type often described in literature (Price 1966, Mattauer 1980) consists of strike parallel joints with one joint set parallel to the fault plane and another oblique to it (Figure 1c). Joint density increases next to the fault. This pattern seems to be independent of the fault type. It is not clear whether these fractures represent genuine mode I fractures, reactivated mode I fractures or genuine shear bands. Another classical pattern (Figure 1d) consists of swarms of joints localized at fault bends or tips, the latter with a more or less marked splaying tendency also called horsetail structure. These joints could be the result of tensile stress concentration either associated with the decrease of the fault slip at fault tips (Rispoli 1982, Granier 1985, Petit and Mattauer 1995) or with fault bends.

3 Joints within patterns of small strike-slip faults

To the best of our knowledge the most complete descriptions of fault-related joints have been given by Rawnsley et al. (1992) based on outcrops in the Bristol Channel area. Figure 2 shows the main characteristics of the perturbed joints in the Nash Point area which is located on the coast of South Wales south of Bridgend (UK map reference SS 92-68). The exposures are in Liassic limestone which can be observed on well exposed layer surfaces. It comprises several decimeter thick, hard, pale grey limestone interbedded with 2 to 10 centimeter thick shale horizons. In this area the regional unperturbed joint pattern consists of parallel joints striking 170° with a mean spacing of about 0.25m. At Nash Point, a series of conjugate strike-slip faults cuts the sequence with trends of about 010° and 160° ±10°. The horizontal fault trace lengths are often greater than the width of exposure (Figure 3a), and they tend to form parallelograms. Due to their continuity and parallelism, it seems likely that the N160° faults of Nash Point could correspond to widely spaced early joints reactivated as faults during the Cenozoic compression.

Joints are compartimented by the parallelograms, so the faults were preexisting to the joints. The latter are always sinuous but with quite regular spacing. Next to faults their general tendency observed over several meters is to be either parallel or perpendicular to the fault trace (Figure 3b). Joint curvature from perpendicular to parallel is often sharp. Locally groups of joints diverging

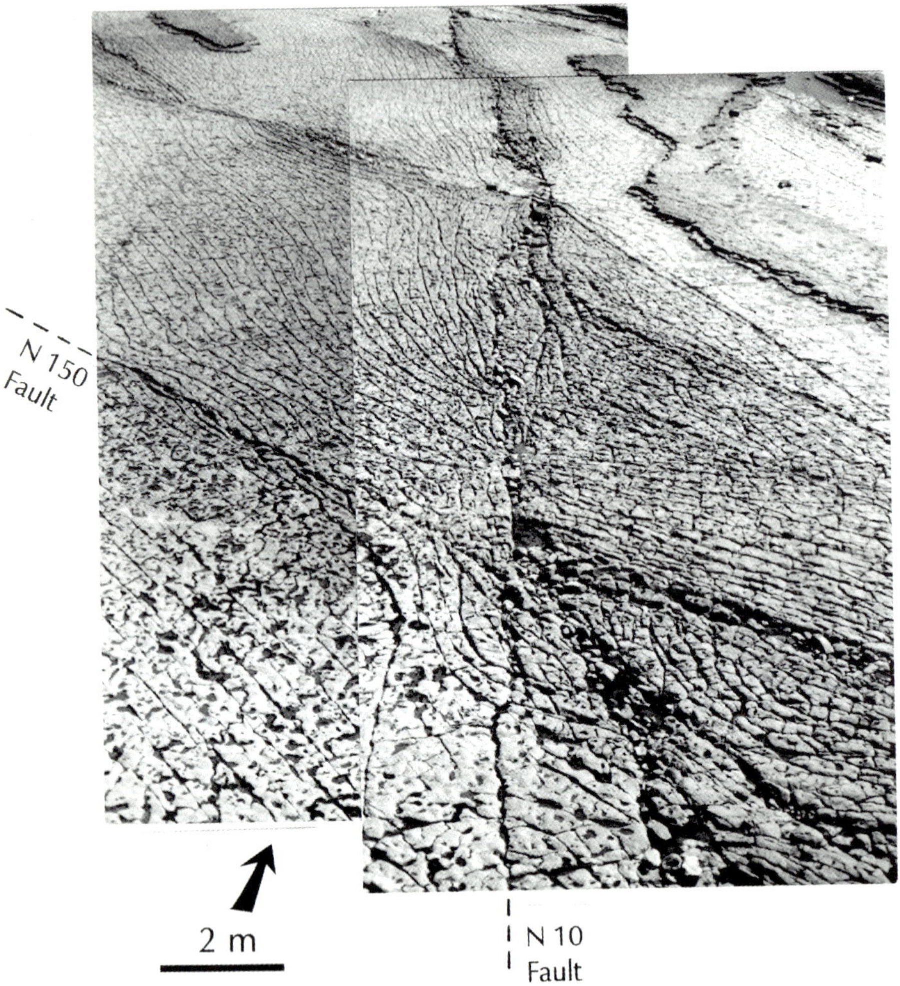

Fig. 2. Oblique view of joint perturbation at small strike-slip faults in the Nash Point area, Bristol Channel. Joint spacing at the front is about 50 cm.

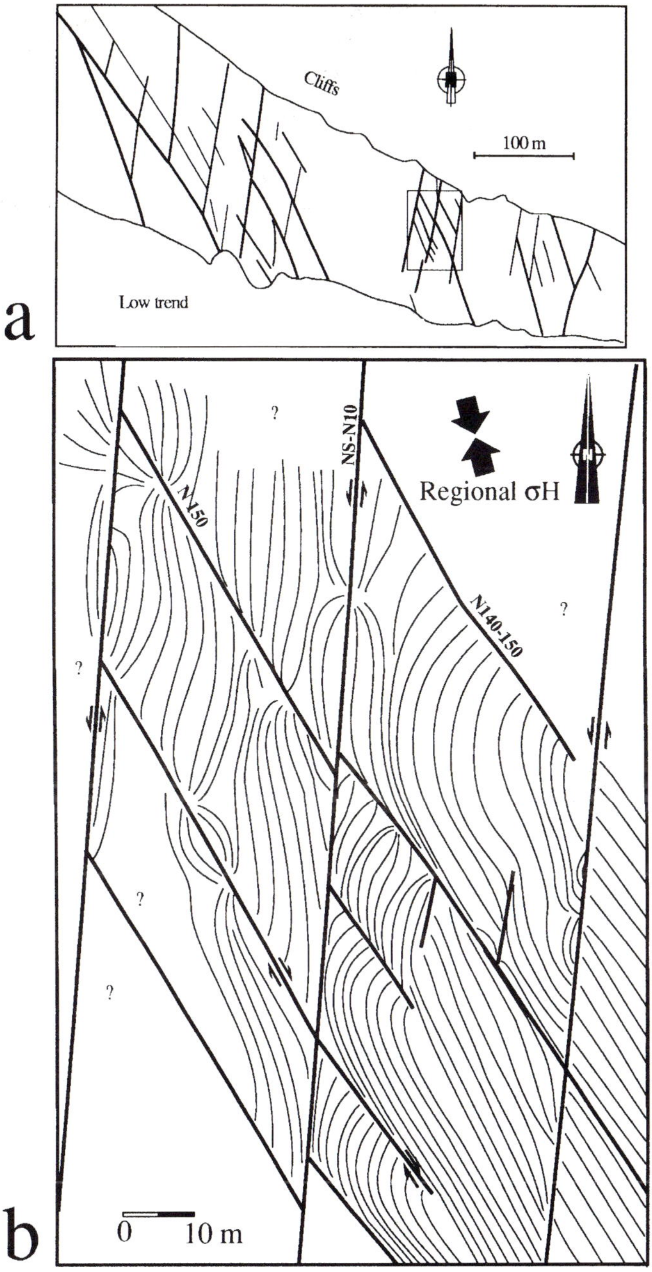

Fig. 3. Sketches of fault and joint distribution in Nash Point area. a - fault distribution along the coast; b - detailed sketch showing the tendencies of joint curvature corresponding to the rectangle in a. Deviation can reach 70° with respect to the N 170° direction of the unperturbed parallel joints away from the fault zone.

away from the faults are observed. ("Diverging" must be taken in a geometrical sense; it does not imply here that joints propagated away from the fault). This striking feature will be further described below. Within the domains bounded by the faults most of the joint deviation is controlled by the longest sides of the boundary faults. Depending on the situation, the curvature within domains can be linked to the fact that joints tend to be parallel to one boundary fault and perpendicular to the other, or to be parallel or perpendicular to both of them. For the latter case the geometry of some individual joints is roughly similar to that of curving-parallel and curving-perpendicular joints which develop at an angle between continuous joints of an older set (Dyer 1988, Rives 1992). There the curved joints are repeated systematically, but this is not observed in the example described here.

All the features described above are on the scale of medium-sized exposures. Larger-scale structures are more difficult to decipher. This is illustrated by the Robin Hood's Bay example on the north Yorkshire coast between Scarborough and Whitby (Rawnsley et al. 1992). In the lower Liassic calcareous and siliceous hard beds interbedded with shales, a joint pattern can be observed on a 200 to 300 meter wide band over more than two kilometers along the coast (grey zone on Figure 4). The pattern is formed of two sets associated in various ways but one

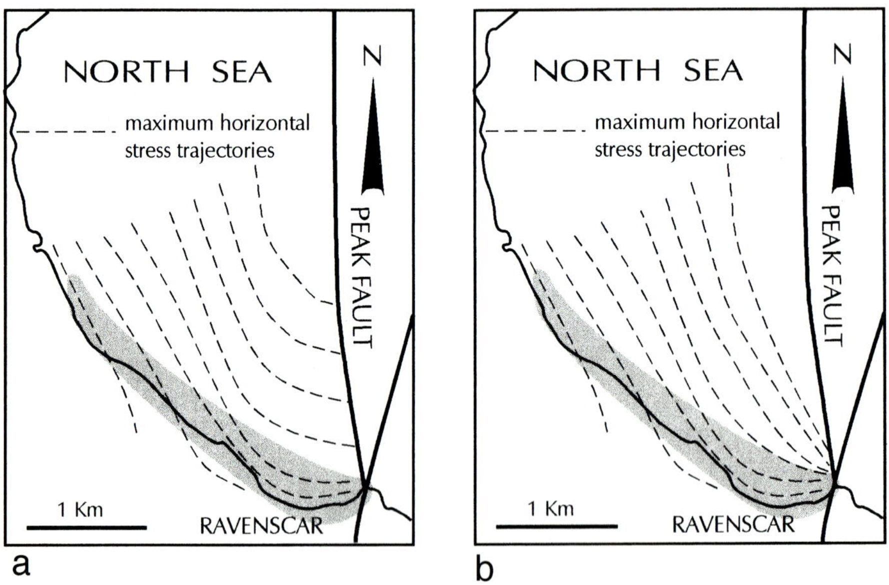

Fig. 4. Two hypotheses explaining joint trajectory deviations observed on a narrow band along the coast in Robin Hood's Bay, Yorkshire, next to Peak Fault. The outcrop zone is grey. a - general curvature of the joints linked to a low friction or open segment on Peak Fault; b - diverging joint model corresponding to a local asperity.

of them shows a progressive curvature from a NE to a nearly EW orientation next to Peak Fault. Due to the elongated shape of the outcrop, two pattern extrapolations are possible leading either to a general curvature model, or to a diverging joint model.

4 Joint divergence from asperities on strike-slip faults

Joint divergence from faults is a striking feature observed in the Bristol Channel area. This structure consists of a number of straight or curved radial joints stemming from narrow zones. The zones of divergence on the faults are often difficult to observe due to the presence of dislocated blocks. When they can be observed they correspond either to a zone of close contact between the two fault walls which can be considered as an asperity on the fault, or to a local fault bend in restraining position as illustrated on Figure 5a. Figure 6b illustrates divergence with curvature. The maximum angle of divergence (defined as the maximum angle between the tangents at joints at a given distance from the fault) is up to 90° in the stemming zone at the fault. It tends to increase away from the fault due to the outward curving of the joints. The diverging pattern can be symmetrical or not with respect to normal to the fault trend at the stemming zone (Figure 7).

When points of divergence are quite close, two basic situations can be observed. The more frequent one is that the two systems of radial joints crosscut each other (Figure 7a). This is illustrated on Figure 6a between point B and C. In a second more rare case it has been observed that systems of continuous, arched joints can link the points of divergence (Figure 7b illustrated by Figure 6a between points A and B). When joints are visible on both sides of the fault, the point of divergence may appear as a center of symmetry (Figure 7d).

One main characteristic of these diverging joints is that the fracture density (i.e. lengths of fracture per unit area) and spacing do not seem to increase next to the stemming point on the fault. Divergence of two adjacent joints is compensated by a new joint inserted between them.

5 Mechanical Interpretation

5.1 Experimental setup

In order to interpret the complex patterns made by the joints, specific experiments have been developed. Industrial brittle varnish, which was very useful for the study of mode I fracture patterns (Rives and Petit,1990 a, b) proved unsuitable for the modelling of fault-related joints because it tends to flake at the point of stress concentration and in shortening conditions. To study joint-fault interaction we used photoelastic models, a well established method of studying the stress distribution in transparent materials which develop birefringence under load (Hetényi 1950; Vishay Micromesures, 1984). Loaded samples are placed between two cross polarizers. Transmitted light forms two kinds of fringes: (i)

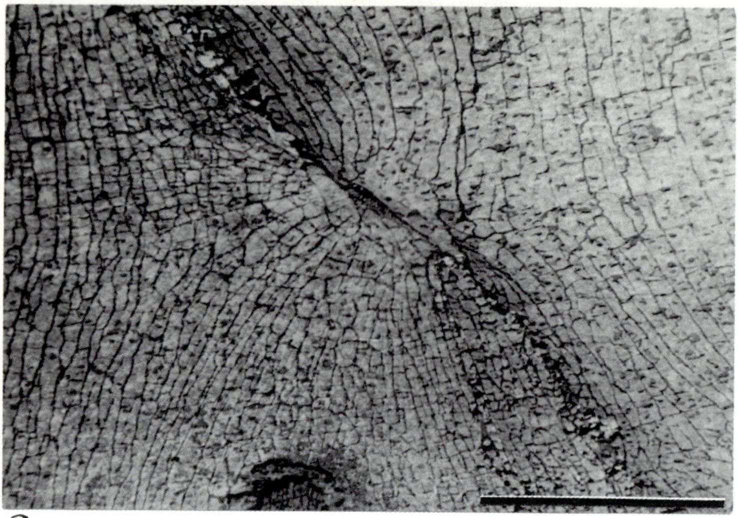

a

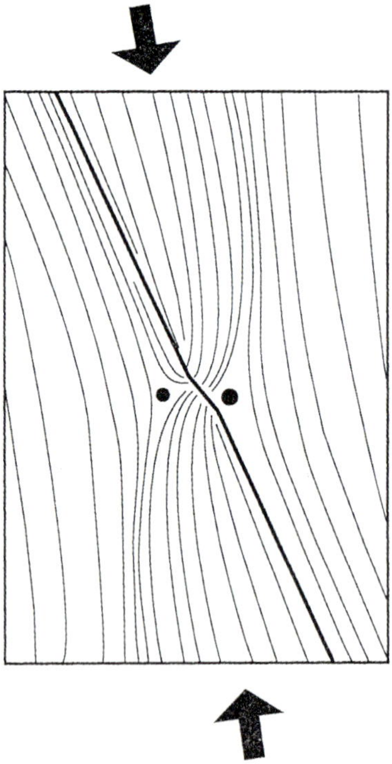

b

Fig. 5. a - Joint deviation at a restraining bend along a N 150° right-lateral small strike-slip fault in the Bristol Channel area; scale bar is about 5m; b - corresponding photoelastic model showing the stress trajectories.

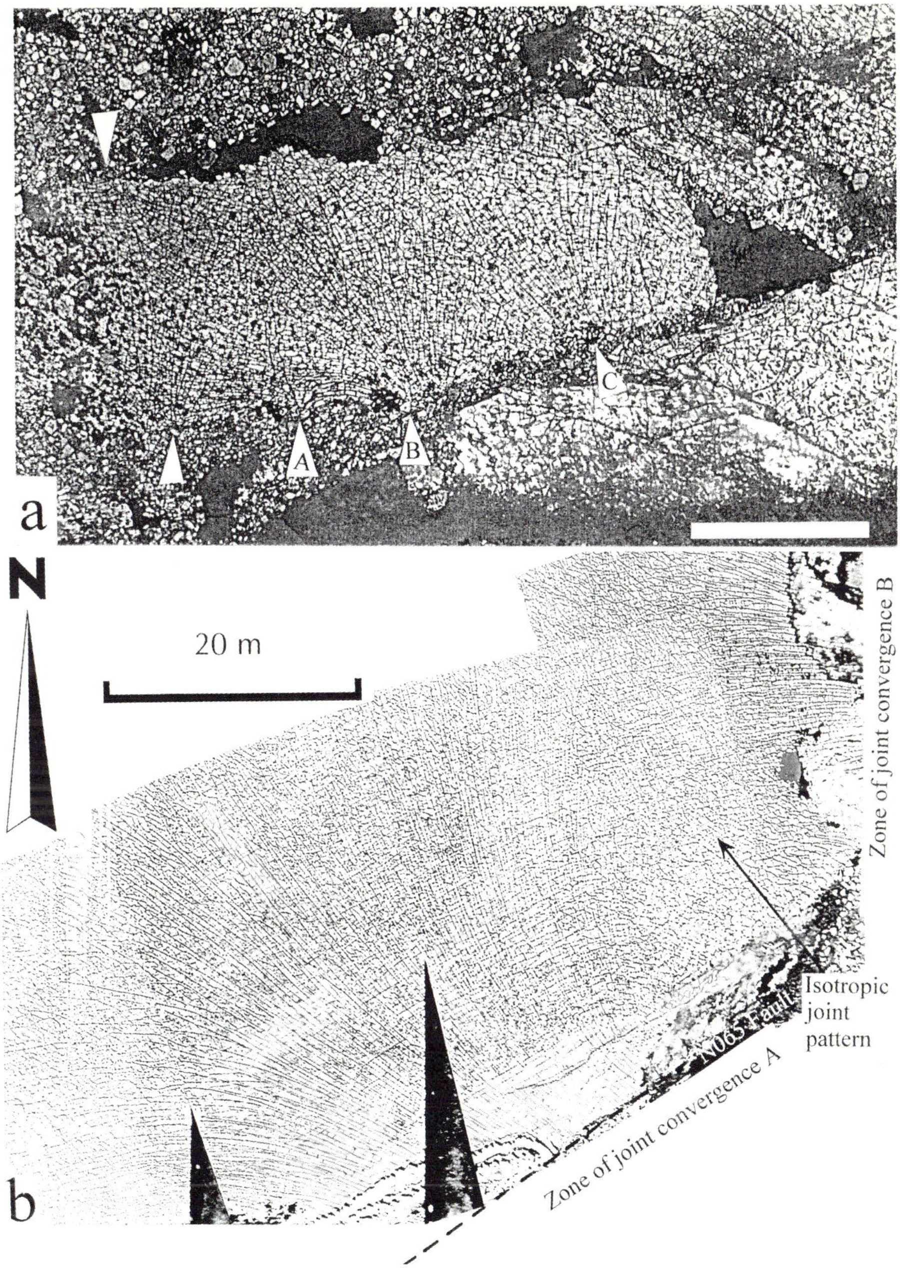

Fig. 6. Helicopter views illustrating diverging joint pattern near faults. a - example from the Nash Point area; triangles indicate the points of divergence aligned along a N140° small right-lateral strike-slip fault; scale bar is about 10m; b: example from Lilstock area (south of Bristol Channel) showing two nearly orthogonal diverging joints patterns coexisting with an isotropic (with no preferred orientation) joint pattern.

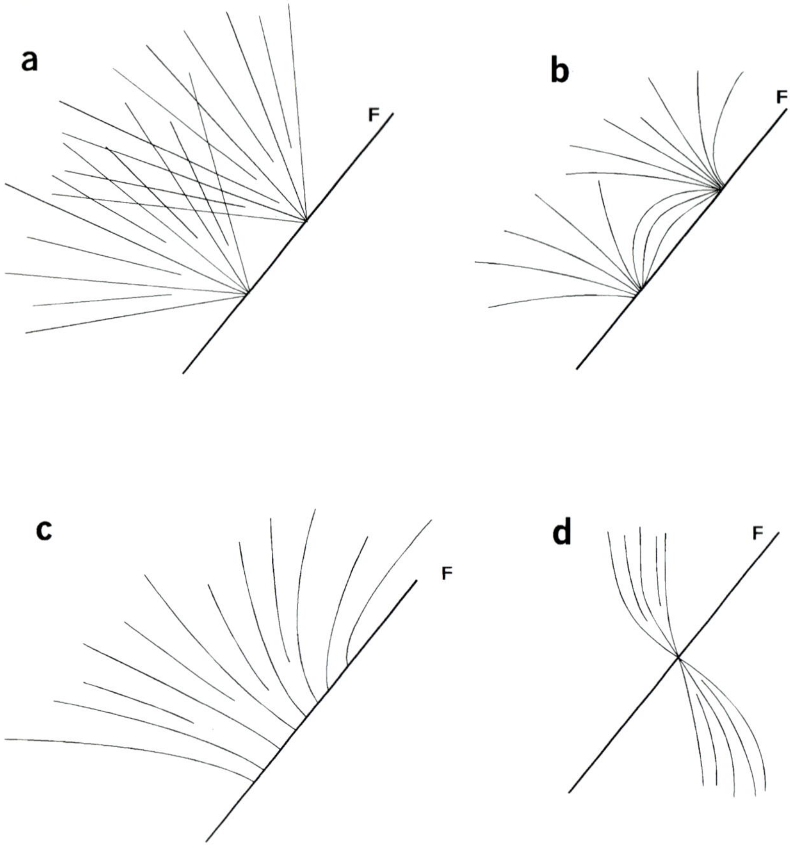

Fig. 7. Various kinds of joint divergence at faults observed in the Lilstock area. a : intersecting symmetrical groups; b : symmetrical groups linked by arched joints; c and d - dissymmetrical groups. In d the point of divergence is a center of symmetry.

the isochromes which correspond to coloured fringes, each of them corresponding to a locus of equal differential stress and (ii) the isoclines which appear as black narrow stripes which are loci of equal angle between the main principal stresses and the loading axis. The changing position of the isoclines when cross polarizers are rotated enables the build-up of the whole pattern of principal stress trajectories unless some isotropic zones are present. The validity of this method (which had already been used by Hyett and Hudson (1990) to study stress concentrations along a sliding fracture) was first verified for an oblique precut defect under uniaxial compression as the stress trajectories and isochromes in photo-elastic and analytical elastic models corresponded very well (Auzias 1995).

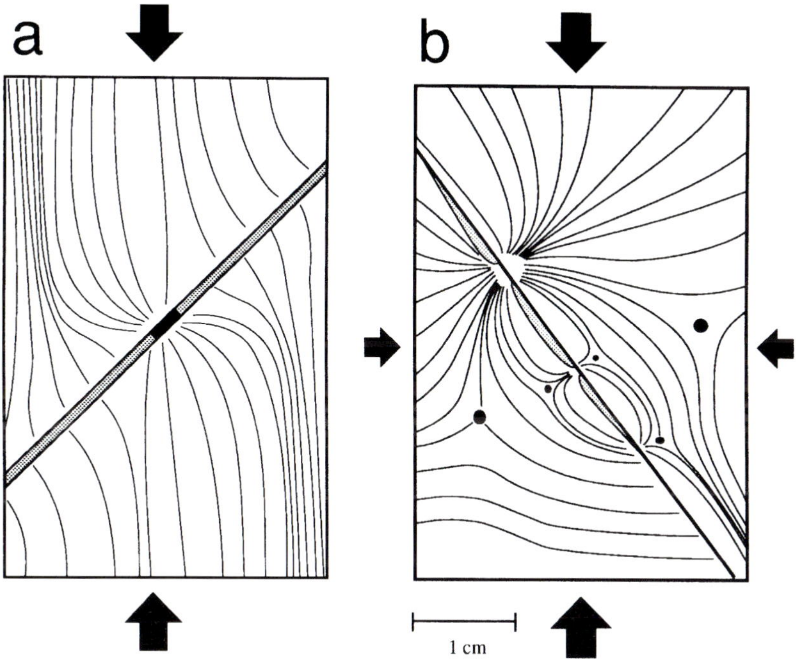

Fig. 8. Photoelastic models of stress trajectories around open faults with imposed asperities. a: single asperity, uniaxial loading; b: multiple contacts with arched stress trajectories, biaxial loading. Black dots correspond to isotropic (zero differential stress) zones where stress trajectories are not defined.

Models of preexisting faults or fault patterns were precut in 5mm thick PMMA plates. This was obtained either by completely sawing the plate and then partially glueing the sawn parts, or by percussion of a blade normal to the plate. In the latter case we obtained a real fracture containing irregularities such as rib marks (Auzias et al. 1993), which gave an excellent model of faults with asperities. The uniaxial loading orientation could be changed at will.

Two main problems were thus studied: the origin of curving-parallel and curving-perpendicular tendencies, and that of the origin of diverging joints. For the latter problem, the test hypothesis was that of stress perturbations linked to an asperity along an otherwise open fracture in a situation which can be compared to that of the Hertzian contact.

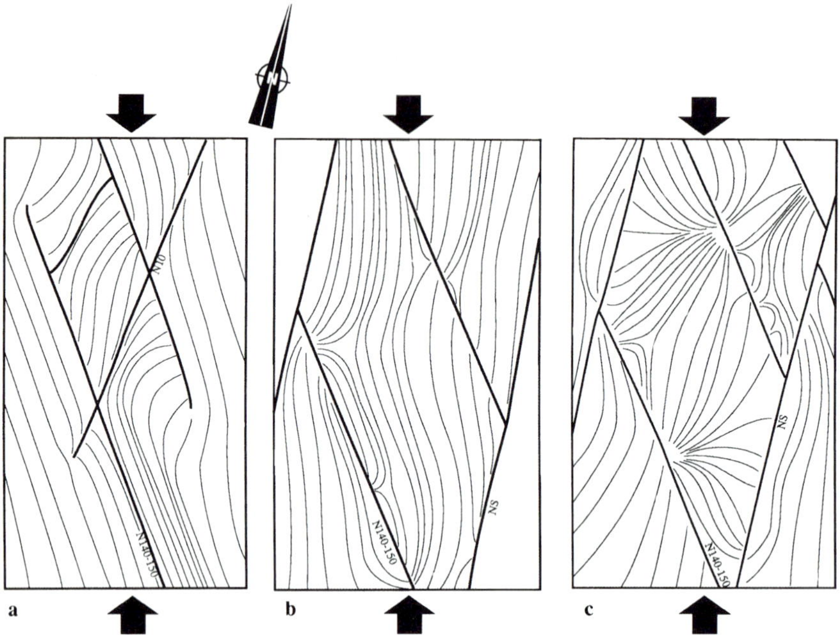

Fig. 9. Photoelastic models of stress trajectory deviations associated with fault configurations observed at Nash Point (Figure 3b). Faults are obtained by percussion of a steel blade on a PMMA plate. Note that the N140-150 fractures are slightly bent.

5.2 Experimental results and interpretation

The simplest experiment was designed to investigate diverging joints and dealt with an open crack with a single imposed asperity. The stress trajectories obtained on Figure 8a have the same geometrical attributes as the observed joints, i.e. divergence from a point on the fault. The result of an experiment where several close asperities were designed along a preexisting precut defect is shown on Figure 8b. The main principal stress trajectories have the same arched shapes as the joints observed in Figure 6a. Figure 5b shows a model with a local change in the fault directions. Stress trajectories are in good agreement with the joint pattern of the restraining bend example on Figure 5a. The model predicts most of the joint pattern geometry but not some radial fractures on the left of the bend. This suggests that joint propagation occurred during two different episodes.

All these experiments clearly confirm the asperity model for joint divergence. Moreover they suggest that when two diverging patterns crosscut each other (Figure 7a) they formed during two different episodes corresponding to two different local stress fields. That means that the first formed pattern was already sealed (i.e. able to transmit tractions) when the second one propagated from the other asperity.

More complex photoelastic models also proved useful in deciphering the origin of the joint patterns associated with the conjugate strike-slip faults. Figure 9 shows three photoelastic models of stress trajectories. These models do not aim to reproduce the observed pattern itself (Figure 3b) but summarize the principal types of geometrical relationships between the NS-N10° and N140°-150° minor strike-slip faults which in some cases form closed parallelograms. The load directions correspond to the regional compression parallel to the non-perturbed N160° joints observed next to but outside the faulted zone. Figure 9a shows the extremely strong influence of fault tips on the curving parallel/perpendicular geometry of joints. This can be compared to the bottom part of Figure 3b. As the joints are practically perpendicular to one side of the fault tip and parallel to the other side, it can be deduced that friction at the fault tip was nearly zero, at least at the level of observation within the layer. In the experiment of Figure 9b, trajectories have a tendency to be parallel to the fault nearly everywhere except at a certain number of points of contact corresponding to local fractographic asperities on the natural fracture plane obtained by percussion. The general tendency within the central parallelogram can be compared to the top left parallelogram in Figure 3b where at least four points of divergence are observed. The experiment of Figure 9c shows a case where the whole pattern is dominated by the influence of two zones of divergence which correspond to very slight changes in orientation on a large proportion of the N140-150 fractures. This case has not been observed on the field, probably because the size of the natural asperity or bend would remain quite small with respect to the lengths of the fault segments limiting the parallelogram. Different loading conditions could be another reason.

5.3 Discussion

All these experiments confirm that joints associated with faults can be excellent markers of paleostress trajectories which in the studied case should correspond to σ_1 since a strike-slip regime is involved. That means that natural joint patterns can provide a precise description of strongly deviated paleostress trajectories at large outcrop scale. The observed joint patterns show the extreme sensitivity of the stress trajectories to the presence of asperities and to large zones of weak friction on faults. This mechanical situation is somewhat surprising as the joints developed in confined conditions. However, weak friction could be the result of interactions between some segments of the fault pattern, or to a generalized contractional or extensional tendency during jointing.

As shown on Figure 4, the geometry is more difficult to decipher on a large scale when outcrops are less continuous. It is *a priori* difficult to decide whether this deviation of one joint set next to the fault can be integrated in a model where all joints become perpendicular to the fault, or whether there is an asperity on the fault. As it seems difficult to assume that a fault can be a free surface over large distances, the asperity model seems more likely.

If the interpretation of joint orientation at faults in terms of stress trajectories seems satisfying, a mechanical paradox must be emphasized. Like any kind of defect in an elastic medium, faults (especially at asperities or at fault tips) are expected to develop stress perturbations. The latter means the coexistence of stress deviations and stress concentrations. In other words, stress deviation cannot occur without stress concentration and vice versa. This basic rule of elastic-brittle mechanics (Jaeger and Cook, 1979; Pollard and Segall 1987) which is currently verified on natural features (Petit and Mattauer 1995) is not respected in the described patterns. If it were, an increasing density of joints closer to the asperity and a decreasing density further away might be observed. Similarly, a great density of joints should be observed on one side of the fault tips due to the tensile stress concentration (Rispoli 1981), whereas joints should

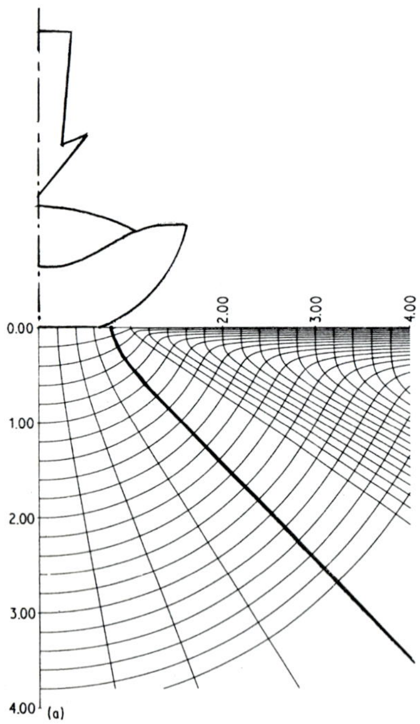

Fig. 10. Hertzian stress trajectories under a spherical punch with the trace of the conical Hertzian fracture (thick black line). Compare with Figure 6b (zone of convergence A)

not be observed on the other side because of the the presence of biaxial compressive stress. Although this point is not quantitatively demonstrated in this short paper, the joint density seems quite constant. In the case of asperities, this paradox is also illustrated by the comparison with the Hertzian contact. The 2D distribution of stress trajectories in this physical model (Figure 10) has characteristics comparable to those observed on the field (Figure 6b). However, in the Hertzian contact, the spatial distribution of tensile stress concentrations determines the formation of one conical fracture (two diverging fractures on an axial section) originating from the immediate periphery of the contact where the maximum tension is present. This situation does not depend much on the shape of the indenter provided it is not sharp (Mouginot and Maugis 1985). On the field examples, no kind of peripheral fractures are observed; instead, fractures have all the trajectory orientations predicted by the model without any noticeable increase in joint density within the set.

This mechanical paradox raises a fundamental question about the mechanical context of jointing. We suggest that (at least in the described examples) the driving force controlling joint propagation has only a slight link with the local stress concentration at asperities or at fault tips, but is mainly determined by the presence of internal contraction. This idea is also suggested by the example of Figure 6b where randomly oriented joints formed between the two groups of diverging joints. Such randomly oriented joints may have propagated with help from an isotropic internal contraction within the limestone layers. Such a contraction, which also explains the weak friction on faults suggested above, could imply the presence of internal tensile residual stress. The latter would appear as the major component of the driving force. In this particular mechanical model there is no more coupling between stress intensities and stress deviations. Indeed the driving force is of internal origin and should be linked to intense and rather homogeneous tensile residual stress. The deviatoric stress which is necessary to define stress trajectories should be small and linked to the tectonic influence. Due to these strong internal tensions, the stress deviations determined by the presence of defects within the layers were not accompanied by significant stress concentrations, which explains the paradox. These ideas are being verified with physical models.

6 Conclusion

This paper is a contribution to the problem of jointing near faults for which even descriptions are rare. The combination of field observations and experiments gives a coherent interpretation of complex joint patterns considered up till now as difficult or even impossible to decipher. In particular photoelastic methods have proved to have great potential for the prediction of joint distribution near faults.

Several important conclusions can be drawn from the examples studied and could be of general interest for the interpretation of joint patterns within the context of layered carbonate rocks. From a fundamental point of view this study confirms that the joints described are unambiguous Mode I fractures i.e. fractures which propagate along the σ_1 stress trajectories; therefore the observed patterns give a record of the complexity of paleostress trajectories around small strike-slip faults during joint formation. As such, the described features can give insight into the *in situ* stress trajectory patterns at depth. The knowledge of such patterns may be important in fractured reservoirs where the *in situ* stress orientations seems vital for the control of the productivity of fluids stored in the fractures.

From a mechanical point of view, the particular case of diverging joints at asperities along faults and of curving-parallel and curving -perpendicular joints at fault tips raises an important problem. These joints cannot simply be interpreted in the light of elastic stress perturbation models. Indeed the stress deviations are not accompanied by fracture concentrations reflecting localised tensions. This paradox could be explained by the dominant role of stored elastic residual stress/strain in the host rock.

Acknowledgments: We are very grateful to Shell research B.V., Elf-Aquitaine and the Institut Français du Pétrole for their interest and financial support. Our thanks to Manuel Willemse for his very helpful suggestions.

References

Auzias V. (1995) Contribution à la caractérisation tectonique des réservoirs fracturés. Thèse de Doctorat, Université Montpellier II, 325p.

Auzias V., Rives. T. and Petit, J.P. (1993) Signification des côtes (rib marks) dans la cinétique de propagation des diaclases : modèle analogique dans le polyméthacrylate de méthyle (PMMA). C.R. Acad. Sci. Paris, t. 317, Série II, p. 705-712.

Dyer R. (1988) Using joint interactions to estimate paleostress ratios. J. Struct. Geol, 10, 685-699.

Granier T. (1985) Origin, damping and pattern of development of faults in granite. Tectonics, 4, N° 7, 721-737.

Hetényi, M. (1950) Handbook of experimental stress analysis. Wiley & Sons Inc., 1077p.

Hyett A.J. and Hudson J.A. (1990) A photoelastic investigation of the stress state close to rock joints, Rock Joints, Barton and Stephansson (eds), ISBN 90 6191 109 5.

Jaeger J.C. and Cook N.G.W. (1979) Fundamentals of rock mechanics, 3rd edition, Chapman and Hall, London.

Mattauer, M. (1978) Les déformations des matériaux de l'écorce terrestre. Hermann édit. Paris.

Mouginot R. and Maugis D. (1985) Fracture Indentation beneath flat and spherical punches. Journal of Materials Sci., 20, 4354-4376.

Petit J.P. and Mattauer M. (1995) Paleostress superimposition deduced from mesoscale structures in limestone: the Matelles exposure, Languedoc, France. J. Struct. Geol., Vol. 17, N° 2, 245-256.

Pollard D. D. and Aydin A. (1988) Progress in understanding jointing over the past century. Geol. Soc. of Amer. Bull. 100, p. 1181-1204.

Pollard, D.D. and Segall, P. (1987) Theoretical displacements and stress near fractures in rocks: with applications to faults, joints, veins, dikes, and solution surfaces. In: Fracture Mechanics of Rock (edited by Atkinson B.K. Academic Press, London. 277-349.

Price N.J. (1966) Fault and Joint Development in brittle and semi-brittle rocks. New York. Pergamon Press, 176 p.

Rawnsley K.D., Rives T., Petit J.P., Hencher S.R and Lumdsen A.C. (1992) Joint development in perturbed stress fields near faults, J. Struct. Geol., 14, 8/9: 939-951.

Rispoli R. (1981) Stress field about strike-slip faults inferred from stylolites and extension fractures. Tectonophysics, 75, T29-T36.

Rives T. (1992) Mécanismes de formation des diaclases dans les roches sédimentaires. Thèse de doctorat., U.S.T.L., Montpellier, 250 p.

Rives, T. and Petit, J.P. (1990a) Experimental study of jointing during cylindrical and non cylindrical folding. Proc. Intern. Conf. Mechanics of Jointed and Faulted Rock, Rossmanith (ed.) : 205-211. Belkema, Rotterdam.

Rives T. and Petit J.P. (1990b) Diaclases et plissements : une approche expérimentale. C.R. Acad. Sci. Paris, t. 310, Série II, 1115-1121.

Tricart P., Blondel T. and Bouaziz S. (1986) Quelques exemples de diaclases précoces en domaine de plate-forme (Tunisie). Leur utilité pour dépister une extension synsédimentaire ou une inversion structurale. C.R. Acad. Sci. Paris, t. 303, Série II, 10, 975-980.

Vishay-micromesures (1984) Encyclopédie d'analyse des contraintes, Micromesures, 533 p.

Dynamic triggering of fault slip

Koji Uenishi and Hans-Peter Rossmanith

Institute of Mechanics, Technical University Vienna, Wiedner Hauptstr. 8-10/325, A-1040 Vienna, Austria

1 Introduction

Earthquakes and rock bursts belong to those natural phenomena which pose a great threat to mankind and they have been associated with mythicism for a long time. In fact, even today with all the computer capabilities man has at his/her disposal there are more questions to be asked than satisfying answers available. In fact, it is not at all clear how earthquakes can be triggered and what kind of connection there exists between an earthquake and the earth's geological structure.

The geologist Gilbert (1884) was the first to notice the immediate aftereffects of the 1872 Owens earthquake in California and as a result of his mapping the terrain he correctly concluded that the elevation of the terrain was produced by repeated sudden ruptures along these faults. Extensive research led to the faulting theory of earthquakes proposed by Reid (Mandl, 1988; Scholz, 1990).

Seismological observations made possible by the worldwide installation of a standardized seismic network have supported the idea of shallow tectonic earthquakes to arise from faulting instabilities and it is the current viewpoint that dynamic faulting is the origin of the majority of seismic events.

2 The model

Fracture mechanics has been particularly advantageous in earthquake modeling where the event is modeled as a dynamically running shear crack. In its simplest form the material is considered linear elastic and the crack (simulating a fault) is cohesionless and the friction term vanishes. A more advanced alternative model is based on stick-slip friction instability along a pre-determined fault surface. This again is a fracture problem but in this crack problem fracture is assumed to occur when the shear stress on the fault reaches the static friction value and the condition for dynamic instability exists. Then, it is assumed that the value of friction will diminish and the shear strength will be decreased in turn.

3 Present study

In this study, our objective is to provide an improved understanding of the interaction between statically pre-stressed surfacing straight fault of finite length and an elastic wave disturbance. The dynamic excitation could be due to a surface blast or due to a remote blast underground. It is assumed that the fault is subjected to initial compression (thrust fault) up to the level such that no slip occurs along the fault due to pre-loading. Then, the loaded fault will be subjected to a dynamic stress along the fault. In general, numerical methods are required to solve wave-fault interaction problems. Therefore, the resulting dynamic stick-slip transition along the fault will be simulated by means of a two-dimensional finite difference technique (SWIFD-FD Program, 1995). The results will be discussed in detail.

We will point out a decisive difference between strike-slip incidence and the ensuing earthquake phenomenon and the interaction between a fault and a stress wave. In the first case the fault develops a dynamic behavior out of static conditions and the ensuing movement is then studied. In the wave-fault interaction phenomenon there is fully dynamic coupling between the dynamic fault and the wave. In other words, the time-dependence of the incident stress wave reflects itself on the magnitude of the shear stresses during slip. Upon shear stress drop due to initiation of fault slip here the normal stress governed by the amplitude of the stress wave will give rise to increasing normal stresses across the fault. As the exact physical connection between normal stress and shear stress during slip is not known for this dynamic situation we assume that during slip the shear stress along the fault will also be allowed to increase.

Two different problems will be investigated: the interaction of a surface blast wave and a plane displacement wave with a fault subjected to compressive prestress. Figure 1 shows the geometry, boundary conditions and the loading arrangement for the numerical simulations. The fault is free to slip if the cohesion stress has been exceeded by the shear stress due to the stress wave.

3.1 Fault activation due to surface blasting

Surface blasting gives rise to P- and S-bulk waves as well as Rayleigh surface waves. These waves interact with the inclined thrust fault which is characterized by a Mohr-Coulomb shear stress criterion, where the coefficient of friction will be lowered to a smaller value upon slip initiation. Figure 1a which pertains to this case shows the basis of the numerical simulation. The dip angle characterizing a wide range of faults from shallow to deeply dipping faults has a pronounced effect on the dynamic behavior of the fault.

Fault interaction with the stress wave is demonstrated in the sequence of three computer generated isochromatic fringe patterns selected to be shown in Figure 2a-c. Figure 2a pertains to the case where the shear wave of the blast impinges on the pre-loaded fault and in the middle section starts to unzip the fault. The radiation of shearing stresses as a result of stress-wave induced unlocking of the fault is seen in Figure 2b. Fault extension is not possible in this model. Figure 2c shows the stage shortly after the diffraction of the Rayleigh surface wave, which hits the acute wedge-shaped section where the fault surfaces. A diffracted Rayleigh surface wave propagating to the right along the free surface can be seen to emerge from the area of interaction. Diffraction also gives rise to a diffracted wave which propagates along the slipped section of the fault. In this case the

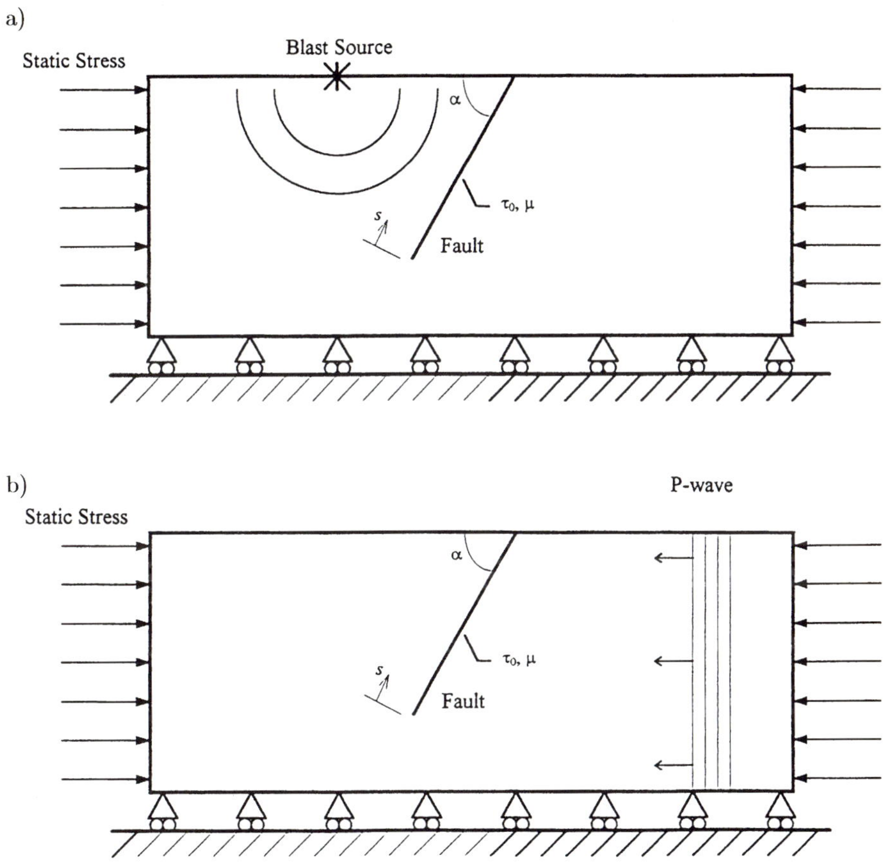

Fig. 1. Geometry and boundary conditions: a) loading of numerical model for surface blast-induced fault slip, b) loading situation associated with wave-induced fault slip of numerical model.

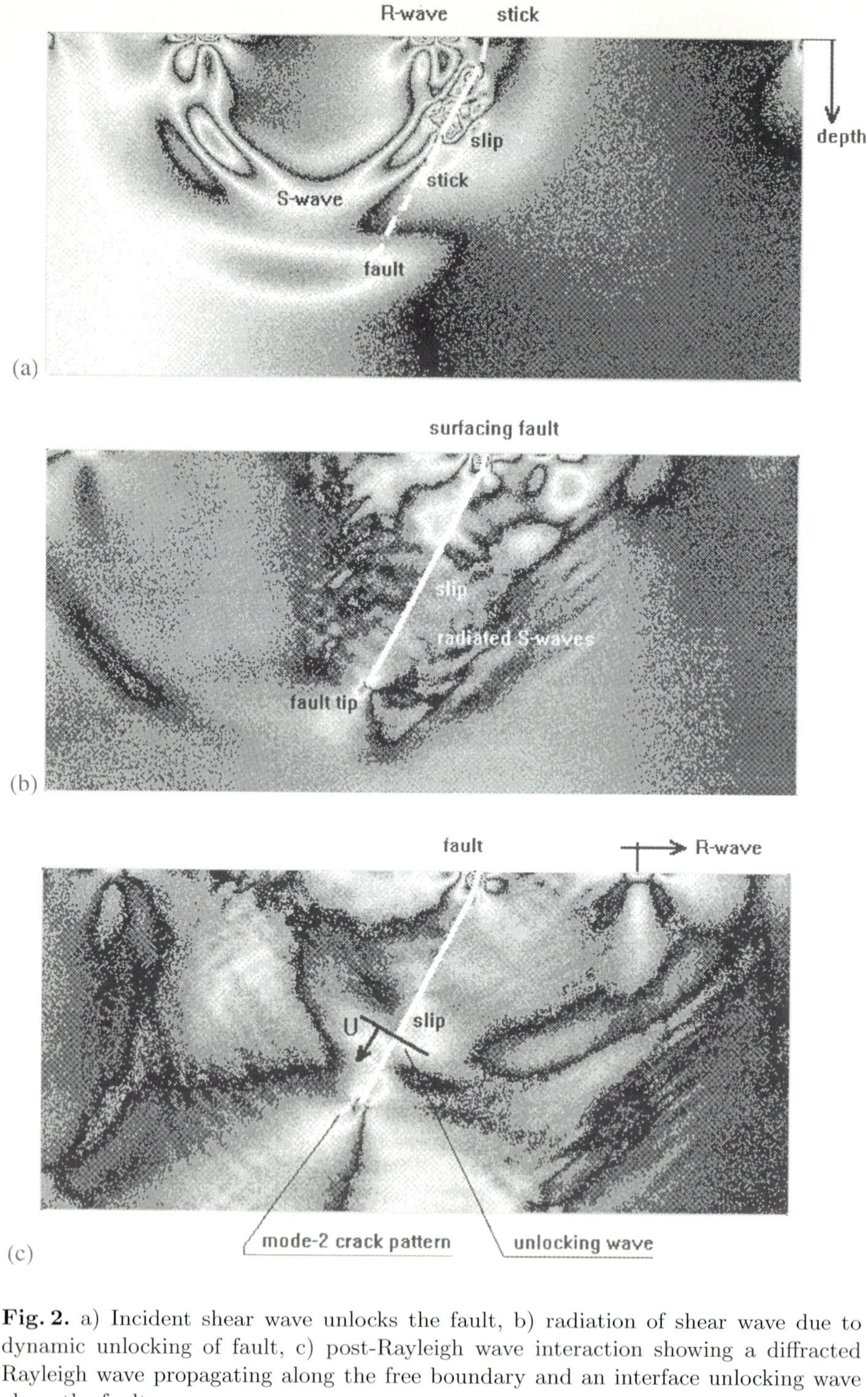

Fig. 2. a) Incident shear wave unlocks the fault, b) radiation of shear wave due to dynamic unlocking of fault, c) post-Rayleigh wave interaction showing a diffracted Rayleigh wave propagating along the free boundary and an interface unlocking wave along the fault.

fault has been unlocked due to the impinging shear wave, however it is possible that the fault will be unlocked or at least re-unlocked by the interface wave which propagates along the fault.

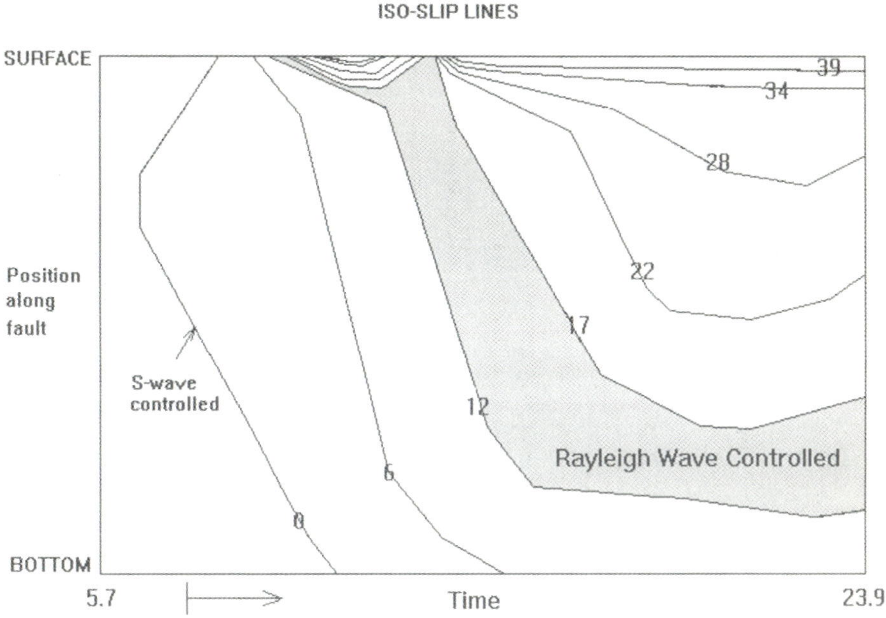

Fig. 3. Interaction of a blast shear wave with an inclined fault showing the dynamics of fault relative slip as a function of time. (Curves pertain to constant values of relative fault slip).

It is educative to study the stick-slip phenomena in a fault position versus time diagram as shown in Figure 3 where lines of constant slip are highlighted. For the geometry given the radiating compressive P-wave of the blast will hit the fault about at halfway through its depth and will compress the fault thus preventing the fault to slip here. However, the inhomogeneous S-wave which is much stronger than the blast P-wave carries its maximum amplitude at a direction of 45°. This shear wave maximum will interact with the fault and will induce slip at the center section of the fault. This slip section will then expand in both directions along the fault. The Rayleigh surface wave controlled regime in the position vs time diagram can clearly be indentified in Figure 3.

3.2 Fault activation due to a plane displacement wave disturbance

If, in shallow underground blasting such as encountered in various mines, a wave disturbance is radiated, this wave may interact with other geological or structural inhomogeneities such as faults and dykes. Fault slip may be induced

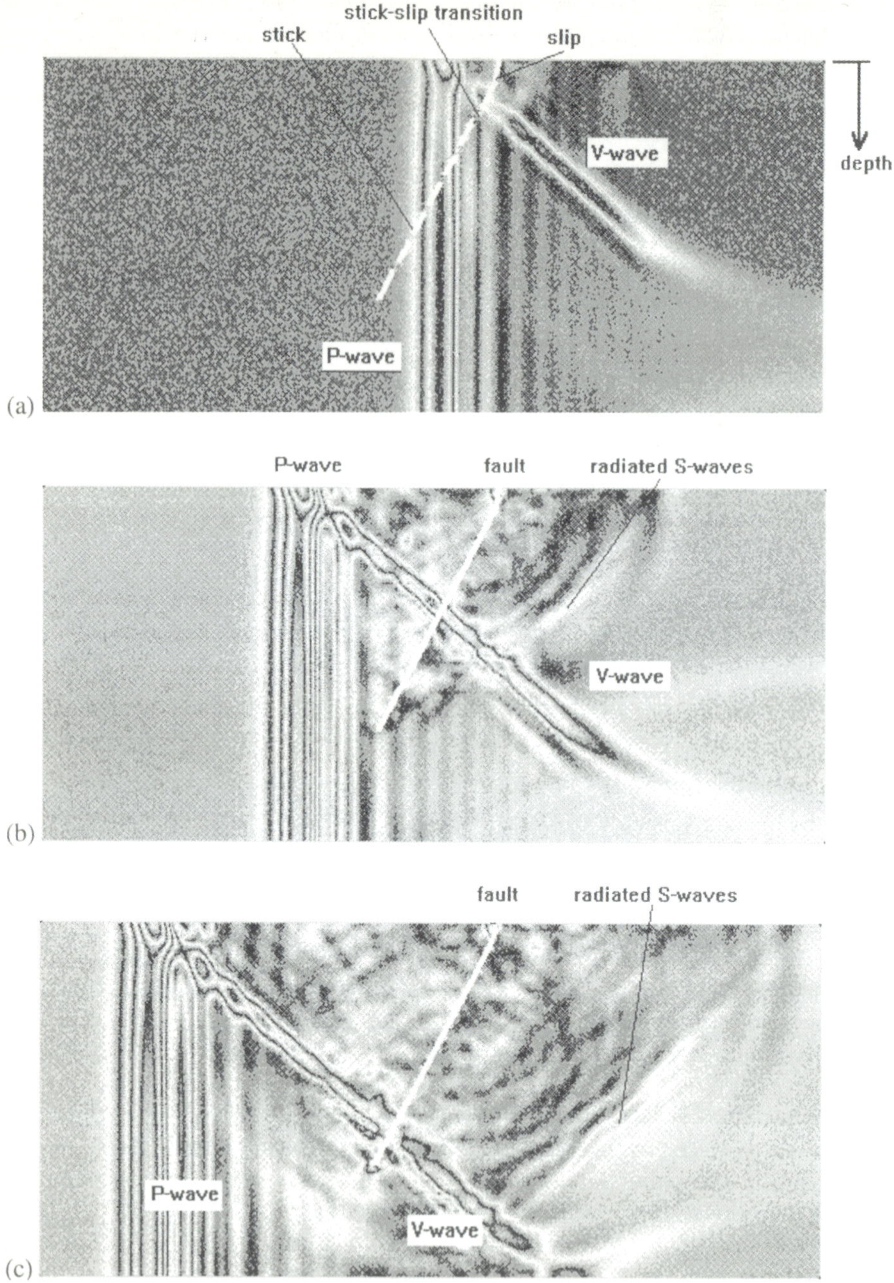

Fig. 4. a) Plane compressive wave and totally reflected von Schmidt wave (V-wave) interact with a fault showing dynamic fault unlocking due to V-shear wave, b) fault unlocking due to V-wave and shear wave radiation from the fault, c) fault unlocking and wave radiation due to V-wave.

and the release of energy stored in the fault will lead to the radiation of seismic or mining-induced seismic waves which in turn may interact with the mining structure. Rock bursts in the mine may result. The phenomenon of mining-induced dynamic triggering of a fault has been observed many times underground together with its devastating effects and threats imposed to mining safety.

Consider the case of the interaction of an inclined fault with a plane displacement wave which impinges on a statically loaded thrust fault as shown in Figure 1b. The horizontal stress acting as a static pre-load is of the order as the maximum dynamic amplitude of the displacement wave.

Figure 4 shows a sequence of three computer generated isochromatic fringe patterns selected to show the interaction of the incident compressive P-wave with the fault. The most striking feature in this case is the emergence of a totally reflected von Schmidt (V-) shear wave due to grazing incidence of the P-wave.

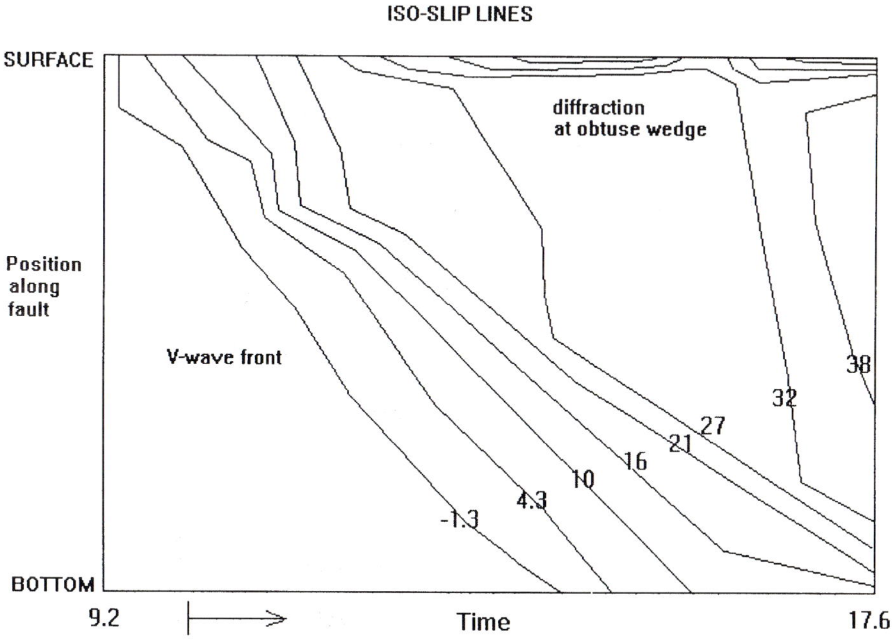

Fig. 5. Interaction of a plane compression wave with an inclined fault showing the dynamics of fault relative slip as a function of time. (Curves pertain to constant values of relative fault slip).

Figure 4a pertains to the phase of the event where part of the P-wave has been transmitted across the squeezed fault and no slip is possible. However the V-wave strongly interacts with and unlocks the pre-loaded fault. This can be seen in Figure 4a. The phase where the V-wave interacts with the fault thereby producing strong radiated S-waves is shown in Figure 4b, and the case where the P-wave has left the fault area and the V-wave is about to hit the fault tip is demonstrated in Figure 4c. This figure also shows a diffracted S-wave which was radiated upon interaction of the P-wave with the intersection of the fault with the surface.

Again, the development of slip areas during the interaction phenomenon can easily be studied by means of the position vs time diagram (Figure 5). The iso-slip lines indicate the movement of the V-wave front along the fault and a rather complex phase of diffraction at the point of surfacing of the fault.

4 Conclusions

To summarize, a finite difference numerical simulation program has been developed to simulate the interaction between a statically pre-loaded thrust fault and a discontinuity wave. One can observe that an impinging stress wave will be scattered about a fault and if the fault is uniformly stressed slip may be induced over certain parts of the fault. This time-dependent slip pattern is strongly controlled by the stress wave pattern. Slip is assumed to occur instantaneously in this model. More complicated shear stress-displacement laws may be employed in a more refined model.

Acknowledgment This work is part of an initial study on numerical simulation of fault dynamics and blasting sponsored by the Austrian National Science Foundation Project No. P10326-GEO. The Finite Difference Code SWIFD has been developed at the Fracture and Photomechanics Laboratory of the Institute of Mechanics at the Technical University Vienna.

References

Gilbert, G.K. (1884). A theory of earthquakes of the Great Basin with a practical application. Am. J. Sci. 27. pp 49-54.

Mandl, G. (1988). Mechanics of tectonic faulting. Elsevier.

Scholz, C.H. (1990). The mechanics of earthquakes and faulting. Cambridge University Press.

SWIFD-FD Program (1995). Fracture and Photomechanics Laboratory, Institute of Mechanics, Technical University Vienna.

Normal fault growth: Evolution of tipline shapes and slip distribution

Emanuel J.M. Willemse[1,2] and David D. Pollard[1]

[1] Rock Fracture Project, Department of Geological and Environmental Sciences, Stanford University, Stanford, CA 94305-2115 USA

[2] now at: EPT-SG, Shell International Exploration & Production, Volmerlaan 8, 2288 GD Rijswijk, The Netherlands. Email: E.J.M.Willemse@siep.shell.com.

Abstract. Descriptions of the 3D geometry of ancient faults suggest that single, continuous normal faults have approximately elliptical tipline shapes, with (sub)horizontal major axes. However, many examples of normal faults are not continuous, but consist of distinct, overstepping segments. Individual segments may have complex shapes as well as a vertical height that exceeds the horizontal length. Where fault segments step laterally, their tiplines are relatively straight and steeply plunging. Tiplines in relays between vertically stepping normal fault segments are also relatively straight, but have a (sub)horizontal attitude. The fault shapes appear strongly related to the presence of neighboring faults and the positions of the fault segments relative to one another.

We evaluate these observations with quasi-static 3D boundary element models consisting of two frictionless, circular normal faults in a homogeneous, linearly-elastic medium, subjected to a uniform dip-slip stress drop. The local slip gradient at the fault tipline is used to calculate the propagation tendency. Mechanical interaction enhances fault propagation at levels where the faults underlap, but impede it where they overlap. For laterally echelon segments, the overlapped portion of the tipline is likely to become straighter, and approach parallelism with the slip vector. Such straight tiplines promote linkage of segments along a continuous line rather than at one or a few points. Also, the slip-parallel, steeply plunging attitude of the line of linkage tends to minimize kinematic constraints during simultaneous slip of adjacent segments. For vertically overlapping segments, horizontal relays are predicted. These relays are perpendicular to the slip vector and thus have a limited preservation potential.

These results suggest that normal faults may consist of a patchwork of segments, linked along lines parallel and perpendicular to the overall slip vector. Each segment in turn is likely to exist of a patchwork of smaller segments. Analysis of the propagation tendency suggests that the envelope of this whole patchwork of linked segments tends to re-establish a smooth elliptical tipline and a smooth distribution of total slip.

1 Introduction

In the last decade, data have become available that document the 3D shape of faults and the slip distribution over entire fault surfaces. The simplest normal faults consist of a single, continuous slip surface bounded by an approximately elliptical tipline, with a subhorizontal major axis (Rippon, 1985; Barnett et al., 1987). Contours of equal displacement discontinuity on such faults can form nearly confocal ellipses around the point of maximum slip which is generally located near the fault centre. However, the majority of normal faults are more complex. For example, many are discontinuous and consist of an array of overstepping segments. Segmented fault geometries can be observed in map view and cross-sectional view; they occur along dip-slip and strike-slip faults, and at length scales that vary between centimetres and kilometres (Gay and Ortlepp, 1979; Aydin and Nur, 1982; Mandl, 1987; Mandl, 1988; Davy, 1993). The tipline of individual fault segments often is far from elliptical in shape. Furthermore, slip variations are complex along segmented faults, especially near geometrical irregularities such as jogs and relays between segments (Peacock, 1991; Peacock and Sanderson, 1991; Peacock and Sanderson, 1994). However, the distribution of total slip along an array of segments often resembles that of a single larger fault with dimensions similar to the array, indicating that segments acting together behave approximately as a single kinematic entity (Walsh and Watterson, 1991).

It is important to investigate the geometry and mechanics of segmented faults both from an applied and a fundamental point of view. Fault segmentation may influence the hydrocarbon trapping potential of faults (Larsen, 1988; Bouvier et al., 1989; Morley et al., 1990; Peacock and Sanderson, 1994) and the scaling between fault length and slip (Gillespie et al., 1992; Dawers and Anders, 1995; Willemse et al., 1996). Furthermore, fault segmentation may affect the process of fault growth. Coalescence and linkage of individual fault segments allows faults to become longer or taller and hence to accommodate more slip (Segall and Pollard, 1983; Kronberg, 1991; Peacock, 1991; Peacock and Sanderson, 1991; Peacock andSanderson, 1994). Consequently, the fault size, 3D shape and slip distribution change through time as individual segments propagate and coalesce. This paper addresses some of the mechanisms that may play an important role in this process.

Research using fracture mechanics principles has shed light on the propagation paths and coalescence of opening-mode fractures such as joints and dikes. This applies to propagation paths of both individual joints due to rotation of the stress field (Cottrell and Rice, 1980), and to propagation and coalescence of a series of mechanically interacting joints (Pollard et al., 1982; Pollard and Aydin, 1984; Olson and Pollard, 1988; Olson and Pollard, 1991; Thomas and Pollard, 1993). However, propagation and coalescence of faults are less well understood. Laboratory experiments suggest that fault propagation is a very complex process, involving localization and subsequent linkage of tensile micro-cracks (Peng and Johnson, 1972; Cox and Scholz, 1988; Petit and Barquins, 1988; Moore and Lockner, 1995; Shen et al., 1995). Nonetheless, studies ignoring such micromechanics have shed light on some of the observations along natural faults, such as the orientation and location of secondary structures (Segall and Pollard, 1980), amount of fault overlap relative to fault separation (Aydin and Schultz, 1990) and propagation paths for a pair of echelon strike-slip faults (Du and Aydin, 1993; Du and Aydin, 1995).

Because segmented faults occur across a wide range of scales, larger segments are likely to form by coalescence of smaller segments. Single continuous faults also may have formed by linkage of initially unconnected segments. This poses several questions. (a) How can many segments with irregular shapes and complex slip distributions coalesce to form a single fault surface with regular tipline and an approximately smooth slip distribution? (b) What mechanisms reduce the complexity in fault shape and slip distribution with increasing linkage? (c) Do the same mechanisms provide coordination between unconnected segments so that all segments together behave as a single kinematic entity? And, (d) how do such mechanisms affect the shape and slip distribution as a fault evolves from a discontinuous segmented array into a larger, more continuous fault surface?

These interrelated questions provided the motivation for this study, suggesting there exist physical mechanisms that promote simplicity and order in fault systems. To investigate this, we first summarize published data on the 3D shapes and slip distributions for single continuous and for segmented normal faults. We then use 3D boundary element models to investigate the relative propagation tendency and slip distribution within an echelon array of unconnected fault segments. The theoretical results are used to infer one possible scenario for the evolution of tipline shapes and slip distribution during propagation and linkage of segments. The resulting model for fault growth through segment linkage is compared with observations from some natural fault zones.

2 Tipline Shapes and Slip Distributions

Outcrop observations show that faults are finite in trace length and bounded by two terminations or tips at which, by definition, the displacement discontinuity decreases to zero. It was not until recently, however, that data have become available to quantify the shape of the fault tiplines in 3D. This can be done by studying a series of maps of the structure at different depths, or alternatively a series of cross sections. Data covering three dimensions come from direct observations in quarries or mines and from interpretations of seismic reflection surveys.

Most mine data on faults come from British coal measures. To maximize recovery, British mining companies have been required since 1911 to keep records of coal measure depths and fault offsets. Understanding fault displacements and the changes in depth of the coal layers around the faults is crucial to predict where the productive measures are located. This legal requirement, combined with the mature state of the mining, has lead to a vast and detailed data set. Because there are many different faulted coal layers, 3D coverage of the faults is obtained. The fault slip distributions and tipline positions are relatively reliable in view of the high precision of the fault throws, which are recorded down to 10 cm (Walsh and Watterson, 1988). Mine data cover faults with offsets up to a few tens of metres and lengths up to approximately one kilometre (Rippon, 1985; Nicol et al., 1996).

Reflection seismic surveys allow the interpretation of larger, kilometre-scale faults. Closely spaced seismic lines provide continuous 3D coverage of a fault or fault zone on a regular grid of data points. However, this technology generally cannot resolve fault throws of less than 10 to 20 metres, so the position of the tipline can only be inferred by extrapolation (Bouvier et al., 1989; Jev et al., 1993). Reflection seismic data may not be able to distinguish very closely spaced faults, but rather would give the impression of one large continuous fault. We will use both mine and seismic data to discuss the shape and slip distributions of single continuous faults, and of segmented normal faults.

2.1 Single Continuous Faults

Two typical examples of single continuous faults are shown in Figure 1. In these strike projections, the fault tipline and the contours of equal slip along the fault are projected onto a vertical plane parellel to the fault strike. The fault tipline and displacement distribution are relatively simple and symmetric. In both cases, the faults are interpreted to be completely bounded by an approximately elliptical tipline. The aspect ratio of well-constrained single faults in the British coal measures varies between 1.25 and 3 (Walsh and Watterson, 1989). The aspect ratio is defined as the horizontal along-strike fault length divided by the down-dip fault height (measured along the fault surface in the dip direction, see Fig. 1a for terminology). Nicol et al. (1996) use reflection seismic data to study normal faults from different regions and lithologies. They observe aspect ratios between 1.0 and 3.4 for single continuous faults and postulate that mechanical anisotropy associated with layering is the main cause for the generally subhorizontal attitude of the major tipline axis.

Muraoka and Kamata (1983) and Rippon (1985) demonstrated that the displacement discontinuity across single isolated faults can vary in a relatively simple and systematic way. Their observations, later corroborated at other locations (Barnett et al., 1987), suggest that the slip on a single fault ideally decreases to zero in all directions from a point of maximum displacement somewhere near the fault's centre. In general, an approximately linear variation is observed. This leads to the nearly equally spaced and approximately concentric contours of displacement discontinuity shown in Figure 1. Because the faults are roughly elliptical in shape, the slip changes more rapidly along the dip direction than along strike.

2D observations show that the slip along some single continuous faults deviates significantly from a linear distribution. For example, the slip distribution may be closely related to layering in the surrounding rock. Across 'stiffer' lithologies, the displacement discontinuity is virtually constant, whilst it may change rapidly in 'softer' lithologies (Muraoka andKamata, 1983; Bürgmann et

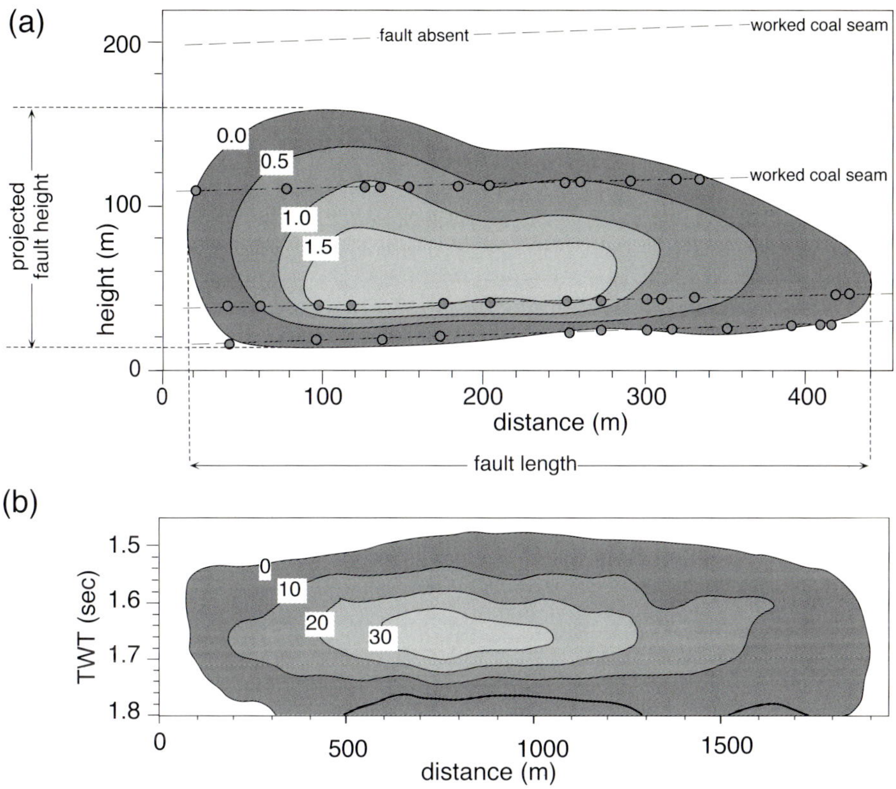

Fig. 1. Shape and slip distribution of single, continuous normal faults. The data are projected onto a vertical plane parallel to fault strike. (a) Contours of fault throw (in meters) based on three completely worked horizons in UK coal measures. The relatively straight lower portion of the tipline is believed to be related to an unusually thick underlying sandstone (after Rippon, 1985). (b) Seismically-mapped slip distribution and tipline shape. Positions in the vertical direction are measured in seconds of seismic two-way-travel times (TWT). Displacement contour values are given in milliseconds TWT and are based on 62 measurements on four horizons. For a seismic velocity of 3 km / s, the maximum slip is about 60 metres, and there is negligble vertical exaggeration. Seismic line spacing approximately 100 meter (modified from Walsh and Watterson, 1991).

al., 1994). Lithologic layering also may affect the shape of the tipline. For example, the straight lower subhorizontal portion of the tipline in Figure 1a is thought to be related to the presence of an underlying, exceptionally thick sandstone unit (Rippon, 1985), probably because of either the stiffness contrasts or the interface between the layers.

2.2 Segmented Faults

The 2D arrangement of the traces of segmented faults as observed in map view or cross-sections has been described for strike-slip faults (Aydin and Nur, 1982; Aydin and Schultz, 1990; Cruikshank et al., 1991; Peacock, 1991), and for normal faults (McGarr et al., 1979; Kronberg, 1991; Peacock and Sanderson, 1991; Peacock andSanderson, 1994). Childs et al. (1995) and Nicol et al. (1996) describe the 3D shape of some normal fault segments. Based on their data, Willemse (1997) distinguishes between discontinuous fault segments that are unconnected in 3D (Fig. 2), and branching / merging faults where multiple fault segments unite laterally or vertically into a single, large fault surface (Fig. 3).

Figure 2a shows a strike projection of underlapping, laterally stepping discontinuous segments. The segments do not have an elliptical shape. Rather, the fault tiplines in the relay zones are remarkably straight and plunge steeply, parallel to the fault slip direction. Throw is approximately constant across each of the three segments, and rapid changes only occur close to the tiplines. The overlapping fault segments in Figure 2b also have steeply plunging tiplines in the relay zone. In this case, slip varies across the fault surfaces and the points of maximum slip are shifted away from the fault centre, toward the relay zone. The 3D shape of both the underlapping and overlapping segments in Figures 2a and b is nearly rectangular. In contrast to single isolated faults, the vertical segment height may exceed the horizontal length. Nicol et al. (1996) report aspect ratios between 0.5 and 1.3 for laterally stepping fault segments.

Figure 2c shows a synsedimentary normal fault that has a distinctly non-elliptical tipline. According to Nicol et al. (1996) the upper, straight, subhorizontal tipline and associated steep displacement gradients are caused by cessation of the synsedimentary fault movement and subsequent deposition of younger layers with no offset (see inset cross-section). The remainder of the fault tipline has a complex, droplet-like shape. This shape, with convex-upwards tipline in the lower, right-hand part of the fault is spatially related to the shape of a nearby fault. The position and shape of the tiplines of both faults are such that the (lateral and / or vertical) overlap between both faults is similar everywhere along the fault surfaces. Where the fault overlap is predominantly in a vertical direction, the tiplines on both faults are relatively straight and subhorizontal, perpendicular to the fault slip direction. Subhorizontal tiplines between vertically stepping normal faults have also been reported by Childs et al. (1995) and Mansfield et al. (1996). Where fault overlap is predominantly lateral, the tiplines generally dip steeply. Although the tiplines of both faults in Figure 2c have a complex shape, the envelope surrounding both faults is relatively smooth and has a simpler shape.

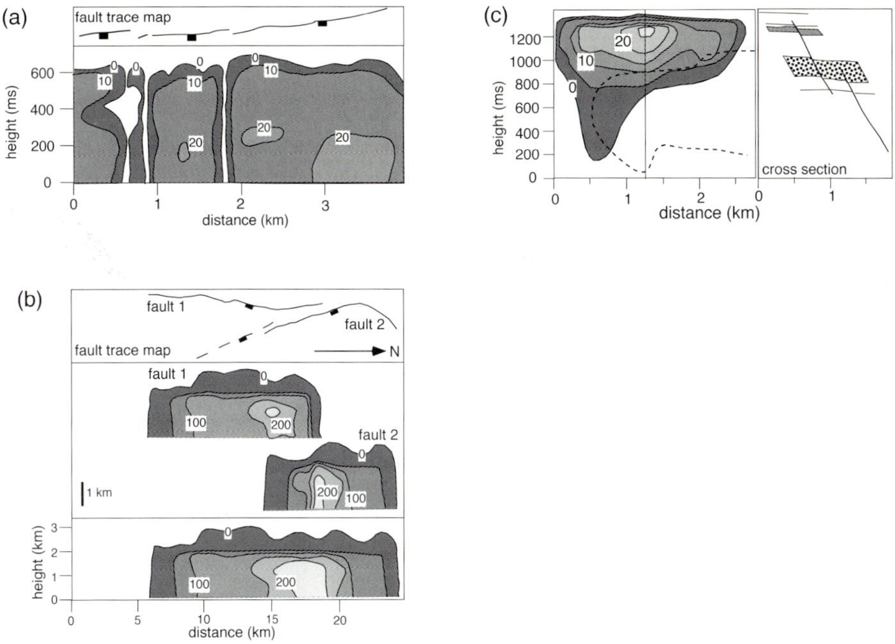

Fig. 2. Shape and slip distribution of discontinuous normal faults. The data are projected onto a vertical plane parallel to fault strike. (a) Fault trace and slip distribution along underlapping fault segments in the Timor Sea. In the relay zone, fault tip lines are straight and plunge steeply. Throw contours are in milliseconds, and based on offsets of 13 horizons mapped on seismic lines spaced every 50 m. Seismic velocity approximately 3 km/s. Vertical exaggeration about 1.4 times. The fault may link downwards at a series of branch points (from Childs et al., 1995) (b) Fault trace map and strike projection of fault slip distribution of two discontinuous overlapping faults (1 and 2) and a minor fault (dashed). Fault tip lines in the relay zone plunge steeply and are relatively straight. Note asymmetric location of area of maximum slip on both faults. Throw distribution on minor fault not shown. Lower slip distribution is aggregate throw on faults 1 and 2. Throw contours are in meters. Seismic lines are E-W and spaced every kilometer. Modified from Childs et al. (1995). Vertical exaggeration about 1.6 times. (c) Cross section and throw contours of synsedimentary fault from the Timor Sea. The fault shape departs significantly from elliptical. The straight upper tipline is related to the end of synsedimentary fault movement, and deposition of the post-faulting sequence. The shape of the remainder of the tip line is mimics the shape of an adjacent overlapping fault (stippled tip line). Throw contours are in milliseconds. Vertical and horizontal scales are approximately equal. Modified from Nicol et al. (1996).

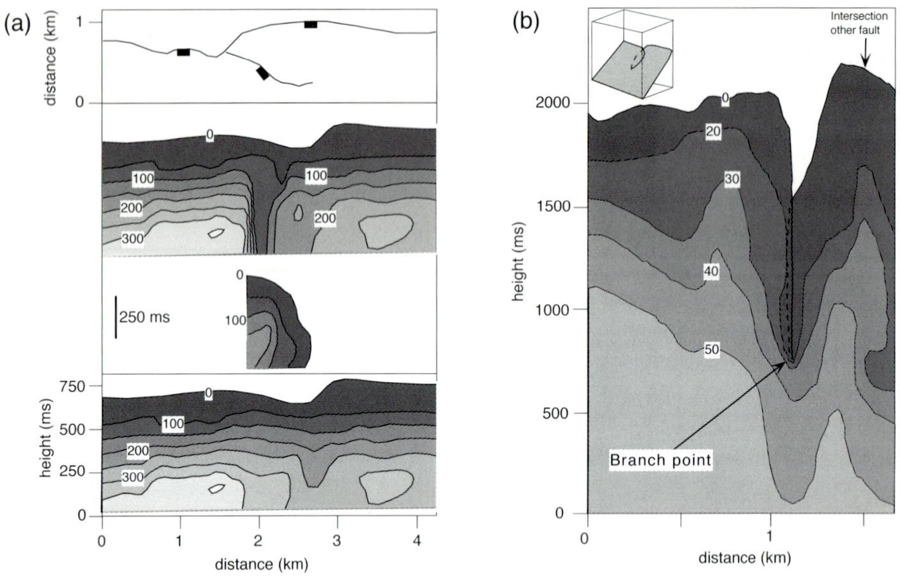

Fig. 3. Shape and slip distribution along branching / merging faults. The data are projected onto a vertical plane parallel to fault strike. (a) Fault trace map and strike projection of fault slip on merging syn-sedimentary normal faults from the Northern North Sea. Interpretation based on 7 reflectors on seismic lines spaced every 100 m. The faults link laterally along a vertical line. Lower slip distribution is aggregate throw on all faults. Contour values in milliseconds. From Childs et al. (1995). (b) 3D geometry and slip distribution along branching / merging fault in the Gulf Coast. The fault surface is continuous at deep levels and divides upwards into two overlapping fault surfaces. It is uncertain whether the faults branch at a point or along a short branchline (see text). Tiplines in the relay zone are straight and plunge steeply. Note deflection of throw contours and steep throw gradients near the fault branch. Irregular contours on right hand side are related to intersection by another normal fault (not shown). Throw contours and depth in milliseconds. Seismic velocity is about 2600 m / s. Seismic lines are spaced every 50 meter. Throw measurements based on offsets of six horizons. Vertical exaggeration about 1.4 times. From Childs et al. (1995).

Figure 3 illustrates the 3D shapes of two merging / branching fault segments. The fault pattern in Figure 3a apparently formed by coalescence of two originally discontinuous segments. Childs et al. (1995) use backstripping to show that, at earlier times, the fault geometry consisted of two overlapping unconnected faults. The reconstructed, unconnected faults share many characteristics of the overlapping faults shown in Figure 2b in terms of tipline shape and off-centre position of maximum slip (see Fig. 13 in Childs et al., 1995). In the present-day configuration, the area of fault intersection is characterized by very closely-spaced, subvertical slip contours. The fault intersection line plunges steeply and is approximately parallel to the fault slip direction. Mansfield et al. (1996) discuss an example from the Gulf Coast where faults merge along subhorizontal intersection lines.

Figure 3b illustrates how a single fault at depth branches upwards into two overlapping segments. In the relay zone, the tiplines of the overlapping segments are very straight, and plunge parallel to the fault dip direction. The segments step to the left and are spaced about 150 metres apart. Because the seismic lines are spaced every 50 metres, it remains uncertain whether the fault branches at a point or along a short (<150 m) branch line. Away from the overlap zone, the slip varies more gradually, except near the intersection line with another fault (see arrow). Close to the branch point, contours of equal slip are more closely spaced, indicating significant increase in slip gradients. Above the branch point, high lateral gradients occur in and to the sides of the overlap zone.

The examples discussed here suggest several common characteristics of segmented normal faults: (i) segmented faults differ significantly from single continuous faults in terms of shape and slip distribution; (ii) the shape of individual fault segments is closely related to the presence of neighboring segments and the spatial position of segments with respect to each other; and (iii) the slip distribution along individual segments generally is asymmetric, and also is related to the spatial position of neighboring segments. Figure 4 schematically summarises these characteristics. Many single continuous normal faults have an elliptical tipline with subhorizontal major axis and a symmetric, near-linear variation in slip (Fig. 4a). The tipline of discontinuous normal fault segments may be elliptical along part of the fault surface, but becomes straight in relays with adjacent faults (Fig. 4b). If the faults step laterally, the lateral tipline plunges steeply and is parallel to the fault slip direction. If the faults step vertically, the top or bottom tipline is subhorizontal and perpendicular to the fault slip direction. Both attitudes can occur along a single fault tipline. Fault segments that are confined at one or both lateral tiplines may become taller than their horizontal length. For branching / merging faults, the intersection lines and relay zone tiplines have similar attitudes (Fig. 4c). The above trends can be recognized along many faults, varying from the seismic scale (Nicol et al., 1995) to sandbox experiments (Childs et al., 1993), and not only along ancient faults but also along active faults (Aki, 1979; Segall and Pollard, 1980).

The fault systematics summarized in Figure 4 motivate an investigation of the physical mechanisms that may influence the shape, slip distribution and

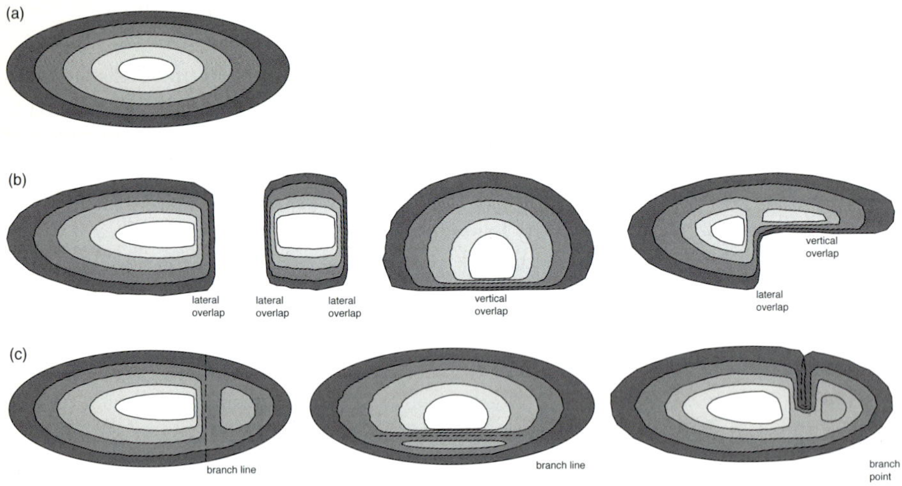

Fig. 4. Idealised fault geometries and slip distributions. (a) Single isolated fault not influenced by neighbouring structures has elliptical tipline and maximum slip at the fault centre. (b) Individual segments from an array of discontinuous faults. Depending on the position of neighbouring faults, the lateral or upper / lower fault terminations may consist of relatively straight tiplines, parallel or perpendicular to the slip vector. Slip distributions are generally asymmetric. (c) Merging faults are characterised by relatively straight lines of linkage that are parallel or perpendicular to the slip vector (left and central sketch). Similarly, relays associated with branching are bound by straight tiplines of similar attitudes.

spatial arrangement of fault segments. One of the mechanisms that is known to influence the slip distribution and to promote certain geometrical configurations is mechanical fault interaction (Segall and Pollard, 1980; Aydin and Schultz, 1990; Pollard et al., 1993; Bürgmann et al., 1994; Willemse, 1997) . Relative movement across one fault segment leads to a local perturbation in the stress field. This perturbation influences the traction and hence the relative movement on neighbouring segments. As an example, Figure 5 shows the perturbation in slip-parallel shear stress around a circular vertical normal fault. The increase in

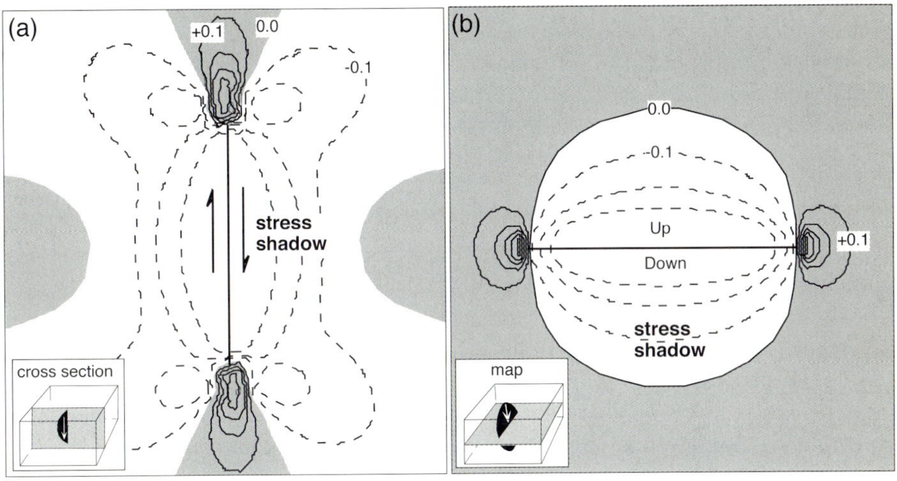

Fig. 5. Contoured distribution of fault-parallel shear stress, acting in the direction of the fault slip vector, about a vertical circular normal fault with a complete, uniform stress drop. (a) Vertical cut through the 3D model provides a cross sectional view in which the stress perturbation is similar, but not identical, to that about a 2D pure mode II crack. (b) Horizontal cut through the 3D model provides a map view, in which the stress perturbation is somewhat like that about a 2D pure mode III crack. The shear stress is enhanced in the shaded areas, and is reduced in the unshaded (stress shadow) areas. Parts of a hypothetical neighbouring parallel fault situated in the shaded areas would locally experience enhanced slip, whilst the slip would locally be inhibited if this neighbour were located in the unshaded areas.

shear stress beyond the tipline of the fault tends to locally enhance the movement on neighbouring faults in that region. Conversely, the decrease in shear stresses to the side of the fault (the "stress shadow") tends to locally reduce slip on neighbouring faults there. Changes in normal stress may also influence slip on neighboring faults: an increase in the normal compressive stress component perpendicular to a neighboring fault surface may increase friction and resistance to slip, leading to potentially reduced slip. Because interaction occurs through local perturbation of the stress field, fault segments can interact without being connected. Willemse et al. (1996) show that interaction among normal fault segments is one possible explanation for some of the first order complexities in the slip distributions described above.

3 Propagation Tendency Related to Fault Slip

In order to determine whether mechanical interaction can explain the observed fault shapes, and the changes of fault shape during fault growth, it is necessary to propose a fault propagation criterion. Given a propagation criterion, one can judge the relative tendency for growth between different fault segments as well as the relative tendency for growth at different locations along the tipline of each segment. Lin and Parmentier (1988) used a fracture toughness criterion to investigate the propagation of a single normal fault in 2D cross-sectional views. Aydin and Schultz (1990) used normalized fault propagation energy to analyse in 2D map-views the enhancing and impeding effects of interaction on the propagation of echelon strike-slip fault segments. They found that mechanical interaction enhances the growth of underlapping strike-slip faults and impedes their growth after some degree of overlap. The effect of interaction on the propagation of normal faults has not been investigated to date. In order to assess the influence of interaction on fault shape, 3D analyses are required to determine the distribution of fault propagation tendency with position on the fault tipline.

According to fracture mechanics principles (Rice, 1968; Rice, 1979; Rudnicki, 1980), the propagation tendency depends on the conditions prevailing near the fault tipline in the so-called near-tip region, $r \ll a$, where a is a characteristic length in the fault plane (Fig. 6a). Here, faults are idealized as cracks and the geometry of the crack tip controls the stress, strain and displacement fields in the near-tip region. In general, for all 3D geometries of cracks embedded in an isotropic, linearly-elastic medium, the near-tip displacement components parallel to the local crack plane are approximated as (Sneddon, 1946; Irwin, 1957; Williams, 1957; Paris and Sih, 1965):

$$u_2 \cong \frac{K_{\mathrm{II}}\sqrt{2r}}{2\mu\sqrt{\pi}}\left\{2[1-\nu]\sin\left(\frac{1}{2}\alpha\right)+\frac{1}{2}\sin(\alpha)\cos(\alpha)\right\} \quad (1)$$

$$u_3 \cong \frac{K_{\mathrm{III}}\sqrt{2r}}{2\mu\sqrt{\pi}}\left\{2\sin\left(\frac{1}{2}\alpha\right)\right\} \quad (2)$$

where K_{II} and K_{III} are the stress intensity factors, μ is the elastic shear modulus, ν is Poisson's ratio, u_2 is the displacement component in the x'_2 direction (perpendicular to the local tipline), u_3 is the displacement in the x'_3 direction (tangential to the local tipline), and r and α are measured in the (x'_1, x'_2)-plane oriented perpendicular to the fault surface as shown in Fig. 6a. The subscripts II and III for K indicate the tipline-perpendicular sliding (i.e. in x'_2) and tipline-parallel tearing (i.e. in x'_3) slip modes respectively. On the face of a 2D crack, these approximations are within $\sim 15\%$ of the exact solution for $r < 0.5a$ and within $\sim 1\%$ for $r < 0.1a$ (Pollard and Segall, 1987) . In what follows, we drop the approximately-equal-to sign, with the understanding that all quantities refer only to these near-tip approximations.

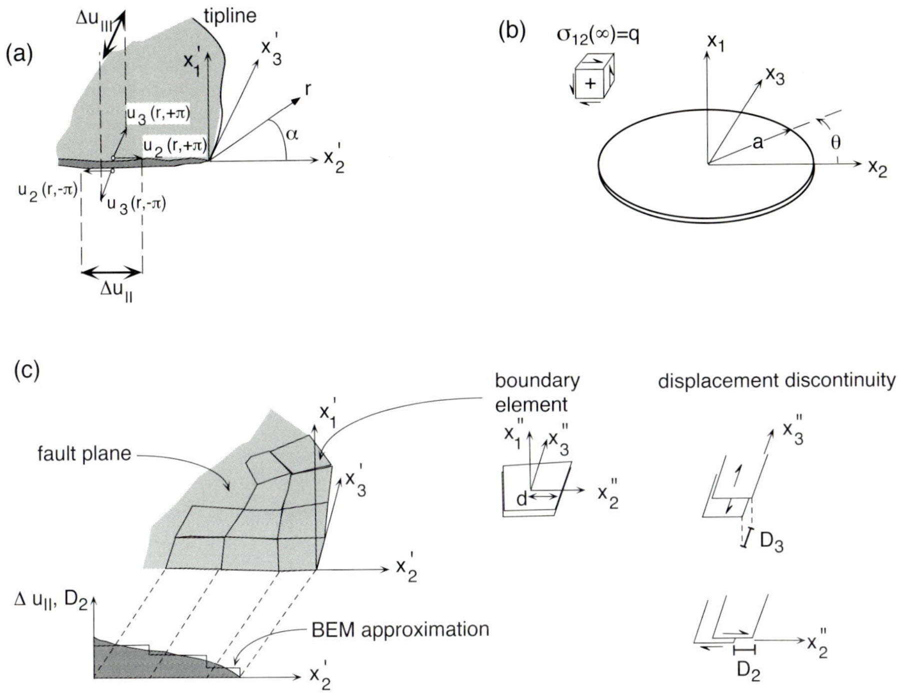

Fig. 6. Figure 6: Fault models based on cracks in elastic materials. (a) Detail of near-tip region of a crack with a curved tipline. The local coordinate system (x_1', x_2', x_3') is related to the local crack plane and tipline attitude with x_1' normal to the crack plane, with x_2' and x_3' in the local crack plane and normal and tangential respectively to the local tipline. (b) Penny-shaped crack of radius a subjected to a uniform shear stress drop q in the x_2 direction. The strengths of the Mode II and III stress intensity factors depend on the angle between the local slip vector and the tipline, and thus vary with location θ on the tipline. (c) Polygonal boundary elements representing part of a crack. Each element has a uniform displacement discontinuity and many elements can be used to approximate the smooth crack displacement discontinuity distribution (shaded) with a stepfunction. The shear displacement discontinuity is composed of a combination of the D_2 (or Δu_{II}) and D3 (or Δu_{III}) components. The stress intensity factors and energy release rate depend on the size of the tip element, measured by the distance d between the tipline and the element midpoint.

The near-tip displacements of points on the two opposing fault sufaces, i.e. for $\alpha = \pm\pi$ (Fig. 6a), are found from (1, 2):

$$u_2 = \frac{\pm(1-\nu)K_{\mathrm{II}}\sqrt{2r}}{\mu\sqrt{\pi}} \tag{3}$$

$$u_3 = \frac{\pm K_{\mathrm{III}}\sqrt{2r}}{\mu\sqrt{\pi}} \tag{4}$$

From these relationships, the near-tip displacement discontinuity (slip) components across the fault are determined:

$$\Delta u_{\mathrm{II}} = u_2(r,+\pi) - u_2(r,-\pi) = \frac{2(1-\nu)K_{\mathrm{II}}\sqrt{2r}}{\mu\sqrt{\pi}} \tag{5}$$

$$\Delta u_{\mathrm{III}} = u_3(r,+\pi) - u_3(r,-\pi) = \frac{2K_{\mathrm{III}}\sqrt{2r}}{\mu\sqrt{\pi}} \tag{6}$$

Note that slip is proportional to the stress intensity factors, and varies as the square root of distance from the fault tip. Re-arranging (5, 6) to solve for the stress intensity factors yields:

$$K_{\mathrm{II}} = \frac{\mu\sqrt{2\pi}}{4(1-\nu)}\frac{\Delta u_{\mathrm{II}}}{\sqrt{r}} \tag{7}$$

$$K_{\mathrm{III}} = \frac{\mu\sqrt{2\pi}}{4}\frac{\Delta u_{\mathrm{III}}}{\sqrt{r}} \tag{8}$$

The stress components in the near-tip region are proportional to the stress intensity factors (Lawn and Wilshaw, 1975), suggesting that these factors are a measure of the deformation in that region. Positive stress intensity factors are associated with positive slip as defined in (5, 6) and illustrated in Figure 6a. Fault propagation is adressed by computing the spatial rate of energy release G, which is a measure of the energy expended as the tipline advances an increment in the (x_2', x_3')-plane, the local plane of the fault. G is obtained from either the stress intensity factors or the displacement discontinuity components (Irwin, 1957; Rice, 1968):

$$G = \frac{(1-\nu)}{2\mu}K_{\mathrm{II}}^2 + \frac{1}{2\mu}K_{\mathrm{III}}^2 = \frac{\pi\mu}{16r}\left\{\frac{1}{1-\nu}(\Delta u_{\mathrm{II}})^2 + (\Delta u_{\mathrm{III}})^2\right\} \tag{9}$$

Because faults are 3D structures, the angle between the local displacement discontinuity and the tipline will generally vary with position along the tipline (Fig. 6b). Therefore the fracture mode at a particular point on the tipline is a combination of mode II sliding and mode III tearing that varies with position along the tipline. For example, for the circular (penny-shaped) crack of radius a shown in Fig. 6b, subject to a uniform remote stress q and no traction on the crack surfaces, the stress intensity factors are (Sih, 1975):

$$K_{\mathrm{II}} = \frac{4\sqrt{\pi a}}{\pi(2-\nu)} q \cos(\theta) \tag{10}$$

$$K_{\mathrm{III}} = \frac{-4\sqrt{\pi a}(1-\nu)}{\pi(2-\nu)} q \sin(\theta) \tag{11}$$

where θ is the angle between the direction in which q acts and the direction of the line connecting the center of the crack to the point on the tipline where the Ks are evaluated. For $\sigma_{12}(\infty) = q$, θ is measured counterclockwise from the x_2-axis (Fig. 6b). K_{II} varies from a maximum where $\theta = 0$ to zero where $\theta = \frac{\pi}{2}$, whilst K_{III} varies from a maximum where $\theta = \frac{\pi}{2}$ to zero where $\theta = 0$. The energy release rate for crack growth consequently varies with position on the tipline.

For simple crack shapes and loading conditions, analytical solutions exist (Sih, 1975; Tada et al., 1985). However, for complex crack geometries or loading conditions, or for multiple cracks that interact with one another, numerical methods are required. We use Poly3d (Thomas, 1993), a computer programme using the Boundary Element Method (BEM) based on displacement discontinuities (Crouch and Starfield, 1983; Becker, 1992). Faults are modeled using a set of polygonal boundary elements, each of which has uniform shear displacement discontinuity components D_2 and D_3 (Fig. 6c). The displacement discontinuity is either prescribed, or in the case of interaction, it is computed to match traction boundary conditions prescribed at the midpoints of each element. Figure 6c shows a representation of the BEM model in the near-tip region that illustrates schematically how polygonal elements can be arranged to approximate the crack surface and tipline. Each component of the displacement discontinuity, D_2 or D_3, is constant over one element, but many elements produce a reasonable approximation to the smooth slip distribution associated with a fault (Fig. 6c).

Equations (7, 8) have been adapted to BEM by multiplying the right hand side by a constant, interpreting r as distance to the midpoint of the tip boundary elements, d, and substituting the uniform displacement discontinuity components of the tip element, D_2 and D_3, for the fault displacements Δu_{II} and Δu_{III} to find:

$$K_{\mathrm{II}} = C\frac{\mu\sqrt{2\pi}}{4(1-\nu)}\frac{D_2}{\sqrt{d}} \tag{12}$$

$$K_{\mathrm{III}} = C\frac{\mu\sqrt{2\pi}}{4}\frac{D_3}{\sqrt{d}} \tag{13}$$

These values are used to compute the energy release rate using (9), which in turn is a measure of the propagation tendency (Aydin and Schultz, 1990). This approach compares very well to analytical solutions, and has an error of less than 5 % even when a 2D crack is divided into only two elements (Olson, 1991) and closely approaches analytical solutions for curved cracks using several tens of elements (Thomas and Pollard, 1993).

When considering the effect on propagation of interaction among multiple fault segments, the relative propagation tendency is found by comparing the calculated G at some point on the tipline to that at the corresponding point on an isolated crack of the same geometry. Based on (9), the relative propagation tendency as a function of position θ on the tipline is given by:

$$\frac{G(\theta)_{\text{interacting}}}{G(\theta)_{\text{isolated}}} = \frac{\frac{1}{1-\nu}\Delta u_{\text{II}}^2(\theta)_{\text{interacting}} + \Delta u_{\text{III}}^2(\theta)_{\text{interacting}}}{\frac{1}{1-\nu}\Delta u_{\text{II}}^2(\theta)_{\text{isolated}} + \Delta u_{\text{III}}^2(\theta)_{\text{isolated}}} \tag{14}$$

Note that the value of Poisson's ratio must be included in the computation of the relative propagation tendency. At any tipline location θ, changes in fault propagation tendency due to interaction can thus be computed if the near-tip displacement discontinuity is known, as well as the displacement discontinuity at the corresponding point on a single isolated fault. In numerical analyses, the computation utilizes D_i rather than Δu_i, and has to be performed carefully as the interacting and isolated fault under consideration must have the same size and shape and must have been modelled using the exact same tip element configuration. Fault growth is enhanced when the propagation tendency is greater than 1.0 and hindered when the propagation tendency is less than 1.0.

This propagation criterion can be compared to other fracture criteria such as those based on the critical energy release rate, G_{crit}, and on fracture toughness, K_{IC} (Lawn andWilshaw, 1975). These two criteria are related to one another through (9) and state that propagation will occur if the energy release rate G or the stress intensity factor K exceed G_{crit} or K_{IC} respectively. The relative propagation tendency criterion as defined here can be made equivalent to these fracture criteria by postulating that the equilibrium solution for the loaded isolated fault is characterized by $G = G_{\text{crit}}$ and $K = K_{IC}$ everywhere along the tipline. The advantage of the relative propagation tendency criterion is that it does not necessitate assigning specific threshold values to G_{crit} or K_{IC}.

The above approach to calculate propagation tendency builds upon the $r^{-\frac{1}{2}}$ stress singularity and the $r^{\frac{1}{2}}$ slip distribution that characterize the near-tip fields of classical elastic fracture mechanics. As r decreases to zero at the crack tip, the stress components theoretically rise to infinity. Although in natural materials such high stresses will be accommodated by inelastic deformation, the above approach is valid provided the size of this zone of inelastic deformation is negligible (also called small-scale yielding) compared to the fracture size (Irwin, 1957). Some opening mode fracture models postulate a zone of high cohesive strength near the tip in order to provide a theoretical limit on the stress components (Dugdale, 1960; Barenblatt, 1962). These concepts have been extended to faulting by postulating a zone of high shear strength or friction near the tipline (Ida, 1972; Palmer and Rice, 1973; Cowie and Scholz, 1992; Bürgmann et al., 1994). In these "cohesive end zone" (CEZ) models the near-tip stress field is dominated by second order terms, and there is no $r^{-\frac{1}{2}}$ increase in stress towards the tip. In contrast to the elliptical slip distribution for faults with uniform strength, the general slip distribution along CEZ faults is generally bell-shaped and can approximate a linear distribution (Bürgmann et al., 1994; Willemse, 1997) . The parabolic slip distribution of (5, 6) is replaced by slip that asymptotically goes to zero at the tipline. For such models, the above criterion for propagation tendency is not applicable.

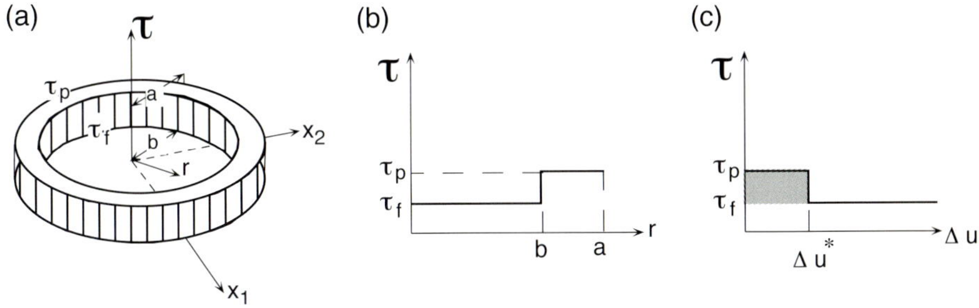

Fig. 7. Penny-shaped crack with cohesive-end-zone. (a) The cohesive-end-zone is modelled as an annular region of greater shear strength τ_p that extends from $b < r < a$ where a is the crack radius. (b) The distribution of shear strength along the fault radius. The central part of the crack ($r < b$) is characterized by a lesser residual shear strength τ_f. The cohesive-end-zone extends between $b < r < a$, and is chareacterised by a greater shear strength τ_p. After Li (1987) (c) Variation of fault shear strength as a function of fault slip. Δu^* is the slip required to reduce the fault strength τ_p to its residual value τ_f. After Li (1987).

Three dimensional examples of such fault models, postulating an annular zone of higher, uniform shear strength τ_p near the fault tipline ($b < r < a$), and a lower, uniform shear strength τ_f on the remaining, central part of the fault surface are discussed by Willemse (1997). An example is shown in Fig. 7. Following 2D analyses by Palmer and Rice (1973), and reviews by Rudnicki (1980) and Li (1987), a fracture criterion appropriate for these boundary conditions is that fault growth occurs when the relative displacement Δu^* at the proximal end of the cohesive zone is sufficient to reduce the fault strength τ_p to the residual value τ_f. This is equivalent to a critical average slip gradient across the CEZ. Using the J-integral, and its equivalence to G (Palmer and Rice, equation 7), a propagation criterion similar to (14) but applicable to CEZ faults can be formulated as:

$$\frac{G(\theta)_{\text{interacting}}}{G(\theta)_{\text{isolated}}} = \frac{\Delta u^*(\theta)_{\text{interacting}}}{\Delta u^*(\theta)_{\text{isolated}}} \tag{15}$$

where $\Delta u^*(\theta)$ is the slip evaluated at or just outside the inner boundary of the cohesive zone (Fig. 7).

3.1 Boundary element model and boundary conditions

As mentioned above, we compute the fault slip distribution with Poly3d, a 3D BEM program based on the displacement discontinuity method and the governing equations of quasi-static linear elasticity theory (Thomas, 1993). The fundamental solution used by Poly3d is that of an angular dislocation in an isotropic linear-elastic half-space (Comninou and Dunders, 1975; Jeyakumaran et al., 1992). The boundary of the half-space represents the traction-free surface of the earth. Several angular dislocations are combined to form a planar polygonal element of constant displacement discontinuity or slip. The planar polygonal element is the 3D equivalent of the straight line segment used in 2D BEM schemes. By joining many polygonal elements, one or more fault surfaces with varying slip can be modelled (Fig. 8a). The faults need not be planar and can have irregular tip-lines.

By prescribing traction boundary conditions at the centre of each polygonal element, mechanical interaction between faults can be modelled. The displacement discontinuity across each element is found by solving a series of linear algebraic equations that describe the influence of each element on all other elements and are consistent with the imposed boundary conditions (Crouch and Starfield, 1983).

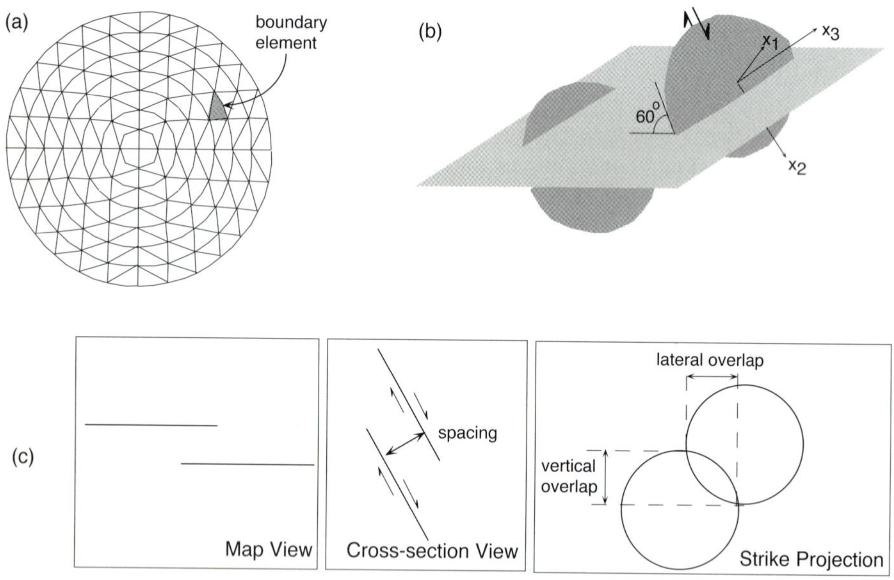

Fig. 8. Model geometry and boundary conditions for the numerical experiments. (a) Typical mesh used to model a fault, consisting of concentric rings of quadrilateral elements. The quadrilateral nature of the elements is visible only near the fault centre. (b) Model geometry, consisting of two deeply buried circular or elliptical normal faults. The faults dip 60° and have only dip-slip movement. The echelon arrangement of the faults may be purely in a lateral direction, purely in a vertical direction, or a combination of both. (c) Model geometry in map view and cross section showing definition of fault spacing and overlap. The definition of lateral and vertical fault overlap is illustrated in the strike projection on the right hand side.

The models discussed below consist of one or more circular normal faults that dip 60°. Multiple faults are parallel to one another and are not linked (Fig. 8b, c). In order to minimize additional complexities due to proximity to the traction-free surface of the earth, the faults are buried deeper than about 10 times their diameter. The faults are modelled in various spatial configurations as described by two parameters, spacing and overlap. Spacing s is defined as the minimum distance between the faults, measured normal to the fault surfaces (Fig. 8c). The distance between the faults observed on a horizontal map view is thus an apparent spacing that is slightly larger than the true spacing (Fig. 8c). The fault spacings considered here range from 0.05 a to 0.25 a, where a is the fault radius. Faults may overlap in both a lateral and vertical direction. The lateral and vertical fault overlaps ol and ov are measured on a projection of the fault tiplines and their definition is illustrated in Figure 8c. The overlaps vary between +2.0 a and -0.5 a, where negative values indicate fault underlap. For faults with centres at different depths the sense of vertical step is such that the faults form an extensional relay (see Fig. 8b). The numerical experiments described here consider pure lateral fault overlap, pure vertical fault overlap, or various combinations.

The faults are loaded remotely by a unit shear stress, σ_{12}^{r} acting in the plane of the faults, and the fault surfaces themselves are free of shear tractions in this direction (Fig. 8b). This results in a uniform stress drop. All other remote shear and normal stresses are zero. The faults accommodate only normal dip slip. Fault normal displacements are presecribed to be zero to prevent opening or interpenetration of the fault walls. Other analyses not described here show that mechanical interaction between normal faults can lead locally to strike-slip components. For the geometries described here, the magnitude of such strike-slip components is at most a few percent of the amount of dip slip. Therefore prescribing the strike slip displacement discontinuities to be zero does not significantly affect the calculated propagation tendency but allows much finer meshes, which are desirable to model CEZ faults, to fit within the computer memory available.

Pure dip-slip motion ensures that the ratio between Δu_{II} and Δu_{III} *at a specific point* near the tipline is the same for an isolated fault and an interacting fault. Because the isolated and interacting fault must be embedded in material with the same Poisson's ratio for proper comparison, the relative strengths of K_{II} and K_{III} are also identical at corresponding points along the tipline. Either of these conditions convenienty reduces (14) to:

$$\frac{G(\theta)_{\text{interacting}}}{G(\theta)_{\text{isolated}}} = \frac{\Delta u_{\mathrm{II}}^2(\theta)_{\text{interacting}}}{\Delta u_{\mathrm{II}}^2(\theta)_{\text{isolated}}} = \frac{\Delta u_{\mathrm{III}}^2(\theta)_{\text{interacting}}}{\Delta u_{\mathrm{III}}^2(\theta)_{\text{isolated}}} \tag{16}$$

Thus, for pure dip slip, the relative propagation tendency *at a specific point* on the tipline can be calculated without considering Poisson's ratio. However, Poisson's ratio influences the 3D variation of the stress components and the strengths of K_{II} and K_{III}. The ratio between K_{II} and K_{III} at a specific point on the tipline for isolated, otherwise identical faults embedded in materials with unlike Poisson's ratio is different, so changes in Poisson's ratio affect the degree of mechanical interaction (Willemse, 1997). Therefore the magnitude of the relative propagation tendency at some given point on the tipline varies for materials with different Poisson's ratio.

Unless stated otherwise, the elastic medium is characterised by a Poisson's ratio of 0.25 and a shear modulus of 15,000 MPa. For both faults with uniform strength (equation 16) and CEZ faults (equation 15), the propagation tendency is independent of the shear modulus because of the normalization by $\Delta u_i(\theta)_{\text{isolated}}$. In some models, Poisson's ratio was set to zero, so that K_{II} and K_{III} have the same maximum value as shown by equation (10, 11).

4 Propagation Tendency Variation along the Tipline

In this section, we compute how fault propagation tendency varies along the fault tipline for various spatial arrangements, fault shapes, and fault strength distributions. 3D analyses for laterally stepping normal faults by Willemse (1997) considered the propagation tendency for a single point on the fault tipline. However, additional insights are gained by investigating how propagation tendency varies with position along the tipline. Mechanical interaction varies with the 3D spatial fault arrangement and with fault shape (Willemse, 1997) and will lead to some non-uniform variation of propagation tendency with position along the tipline. For example, the tiplines of two laterally-stepping circular faults may overlap at some depth levels, decreasing propagation tendency locally, but these same tiplines underlap at other depth levels, enhancing propagation tendency locally. This can be appreciated by viewing the fault surfaces in a strike projection as shown in the inset in Figure 9a.

We begin with the analysis of two circular, laterally stepping faults ($ov = 2.0a$, $-0.2a < ol < 0.5a$) subjected to a uniform stress drop. In this case the distribution of propagation tendency is symmetrical about $\theta = 0°$. For underlapping faults (i.e. $ol = -0.2a$ and $ol = 0.0a$, Fig. 9a), the propagation tendency increases everywhere along the tipline, especially on the relay side of the fault ($-90° < q < 90°$). Furthermore, the propagation tendency distribution has a single maximum which occurs where the underlapping faults come closest together in 3D ($\theta = 0°$). The magnitude of the propagation tendency increases as the distance between the faults decreases: faults with zero overlap have a greater propagation tendency than faults that underlap. The jagged appearance of the curves in Figs. 9, 10, and 12 is caused by numerical imprecision and should be ignored.

For overlapping faults, the variation of propagation tendency with θ is quite different. The propagation tendency is locally reduced where the faults overlap. This leads to two maxima of the propagation tendency separated by an arc on the tipline of lesser propagation tendency (Fig. 9a). For significant fault overlap ($ol = 0.2a$ and $ol = 0.5a$), propagation is inhibited locally (values less than 1) for low values of θ, but enhanced at other positions along the tipline where the faults underlap.

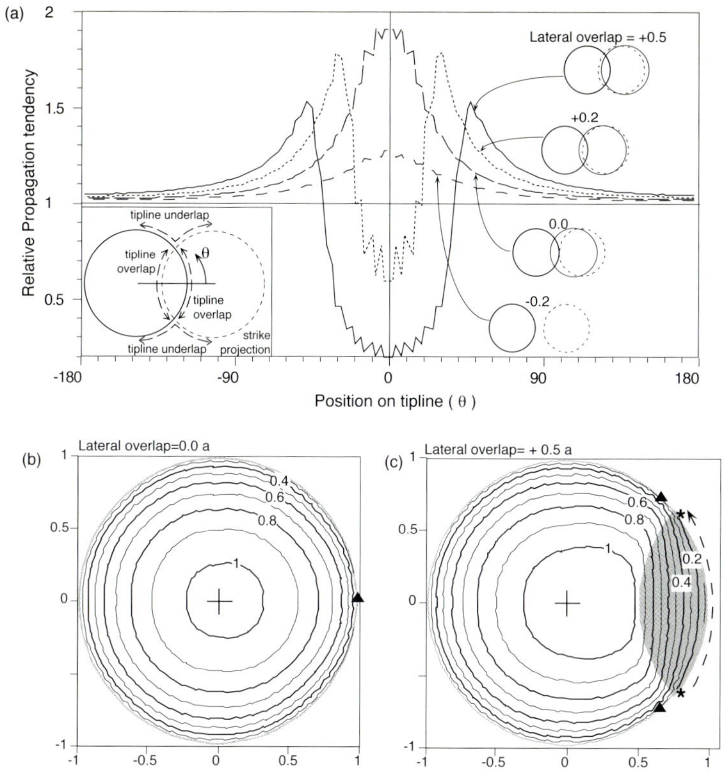

Fig. 9. Propagation tendency and slip along circular faults for various lateral fault overlaps. The faults are subjected to a uniform stress drop and spaced at 0.05 a where a is the fault radius, and embedded in material with Poisson's ratio equal to 0.25. (a) Variation of propagation tendency along the tipline of the left fault (see inset). Underlapping faults (short and long dashed curves) have increased propagation tendency everywhere, and especially where the fault tiplines are close together (low values of θ). Overlapping faults (dotted and solid curves) are characterized by increased propagation tendency along most parts of the tipline, but a reduced propagation tendency where the faults overlap (low values of θ). (b) Contours of slip on the left fault for the case of zero lateral overlap. A contour value of 1 indicates the maximum slip that would occur on a single isolated fault. The contours are slightly asymmetrical but still resemble the concentric circular contours of an isolated fault. The magnitude of the maximum slip increases, and is located to the right of the fault centre (indicated by cross), i.e. towards the adjacent fault. Slip gradients are steeper near the right-hand side of the fault. The solid triangle indicates the point of maximum slip in the near tip region and hence maximum propagation tendency. Compare to dashed curves in (a). (c) Contours of slip on the left fault for the case of 0.5 a lateral overlap. The contour pattern deviates significantly from the circular contours of an isolated fault. The shading in the right hand diagram outlines the area of overlap with the neighbouring fault. Across most of this region, the near-tip slip is reduced. Stars indicate points where the propagation tendency changes from enhanced to reduced. The solid triangles indicate points of maximum propagation tendency.

These variations in propagation tendency can be further appreciated by considering the 3D slip distributions on the fault plane. Figures 9b and 9c show contours of equal fault slip on the left-hand fault, where fault slip has been normalised to the maximum slip that would occur on an isolated fault. Figure 9b shows the slip contours for the case of zero overlap. The greatest slip exceeds that on the isolated fault and the contours show that the position of maximum slip is located just off-centre towards the adjacent fault. Also, the outermost slip contour is deflected slightly towards the tipline. These asymmetries in the slip distribution indicate that the near-tip displacements are greater on the right-hand side of the fault, maximal at $\theta = 0°$ (marked with solid triangle in Fig. 9b), and decay gradually away from this point toward the left hand side of the fault.

For overlapping faults (Fig. 9c), the magnitude of the maximum slip is greater and occurs even closer to the adjacent fault. However, the slip distribution in the relay zone is more complex. In the area of fault overlap (shaded in Fig. 9c), the contours of fault slip depart significantly from the concentric circular pattern of an isolated fault. In this region, the contours are deflected away from the tipline indicating a decrease in the near-tip displacements. In the region of overlap, the slip contours are relatively straight and approximately equally spaced, indicating a nearly linear change in fault slip. Relative to the elliptical slip distribution along isolated faults, the slip near the tip has decreased across most of the region of overlap, resulting in a locally decreased propagation tendency. However, outside the region of fault overlap, the slip is greater near the tipline compared to that on an isolated fault, resulting in a locally enhanced propagation tendency. The transition between enhanced and reduced propagation tendency (stars in Fig. 9c) occurs just inside the region of fault overlap. The points of maximum propagation tendency are located just outside the region of fault overlap (solid triangles in Fig. 9c).

Similar analyses of propagation tendency have been done for purely vertically stepping faults. In this case the propagation tendency is symmetrical about $\theta = 90°$ (see below). For vertically-stepping faults, the mechanical behavior is more strongly influenced by Mode II fault interaction, whereas laterally-stepping faults have a greater Mode III contribution. However, the general trends in computed propagation tendency as based on the energy release rate through (16) are similar to those of the laterally stepping faults. The actual values are slightly different because of the different spatial variations in the shear stress component that drives the slip (cf. Fig. 5a and 5b) and the different dependence of the Mode II and III on Poisson's ratio (7, 8).

Finally, the effect of obliquely-stepping faults that are offset both laterally and vertically has been investigated. Figure 10 shows the response for two circular faults that overlap vertically by $0.5a$, for various magnitudes of lateral fault overlap ($0.0a < ol < 2.0a$). To facilitate interpretation of the results, we first discuss models for which Poisson's ratio is zero, so that the mode II and III stress intensity factors have equal weight in the calculation of G. The variations in propagation tendency along the fault tipline therefore are controlled only by spatial fault arrangement and changes of the $\frac{K_{II}}{K_{III}}$ ratio with θ (Fig. 10a). The propagation tendency varies approximately symmetrically if the fault underlaps in 3D ($ol = 0.0a$ and $ol = 0.5a$). There is only one maximum in the propagation tendency, which occurs where the tiplines come closest together. The propagation tendency decays nearly symmetrically away from this point.

The position θ of the point of maximum propagation tendency changes with increasing lateral fault overlap. As the faults approach one another, a zone of overlap forms locally. In this zone, the propagation tendency is reduced. This leads to two local maxima in propagation tendency, separated by a trough, similar to what was observed for pure lateral overlap (Fig. 9) and for pure vertical overlap. In this case, however, the two maxima have different magnitudes, and the variation of propagation tendency along the tipline is generally asymmetric. This asymmetry directly reflects the asymmetric fault arrangement.

Figure 10a suggests that the greatest propagation tendencies occur on the Mode II dominated upper part of the left fault (i.e. $\theta \sim 90°$). For the right fault the greatest tendency occurs for $\theta \sim -90°$, i.e. along the Mode II dominated lower part of the fault. The degree of asymmetry changes with increasing lateral fault overlap. In the end-member case where the faults overlap completely in a lateral direction (i.e. lateral overlap $ol = 2.0a$), there is only a vertical step and, as discussed above, the distribution of propagation tendency re-establishes symmetry about $\theta = 90°$.

For non-zero Poisson's ratios, which are more representative of rocks, the interaction is more complex because the strengths the mode II and III stress intensity factors are not equal. This may either diminish or strengthen the degree of asymmetry, depending on the exact fault arrangement. For example, for the geometries considered here and a Poisson's ratio of 0.25, the distribution is nearly symmetric as shown in Fig. 10b. In this case, the propagation tendency is slightly greater on the Mode III dominated side of the left fault (i.e. for low values of θ).

All the above analyses consider circular faults with a uniform stress drop. Faults with other shapes and CEZ faults were also investigated. Although the intensity of interaction depends on the size of the cohesive zone and to a lesser extend on fault shape (Willemse, 1997), resulting in slightly different distributions of propagation tendency, the overall trends are similar to those described above.

5 Evolution of Tipline Shape

Our model results show how the propagation tendency of normal faults is influenced by mechanical interaction between neighboring segments. The effect on propagation tendency varies along the fault tipline, depending on the spatial arrangement of the faults. Based on this, we *infer* how faults might propagate and how this would affect their tipline shape. Figure 11 shows a schematic and conceptual model of how faults may propagate based on these results. Laterally-stepping and underlapping faults have increased propagation tendency everywhere along their tipline, especially where their tiplines are adjacent (Fig. 11a).

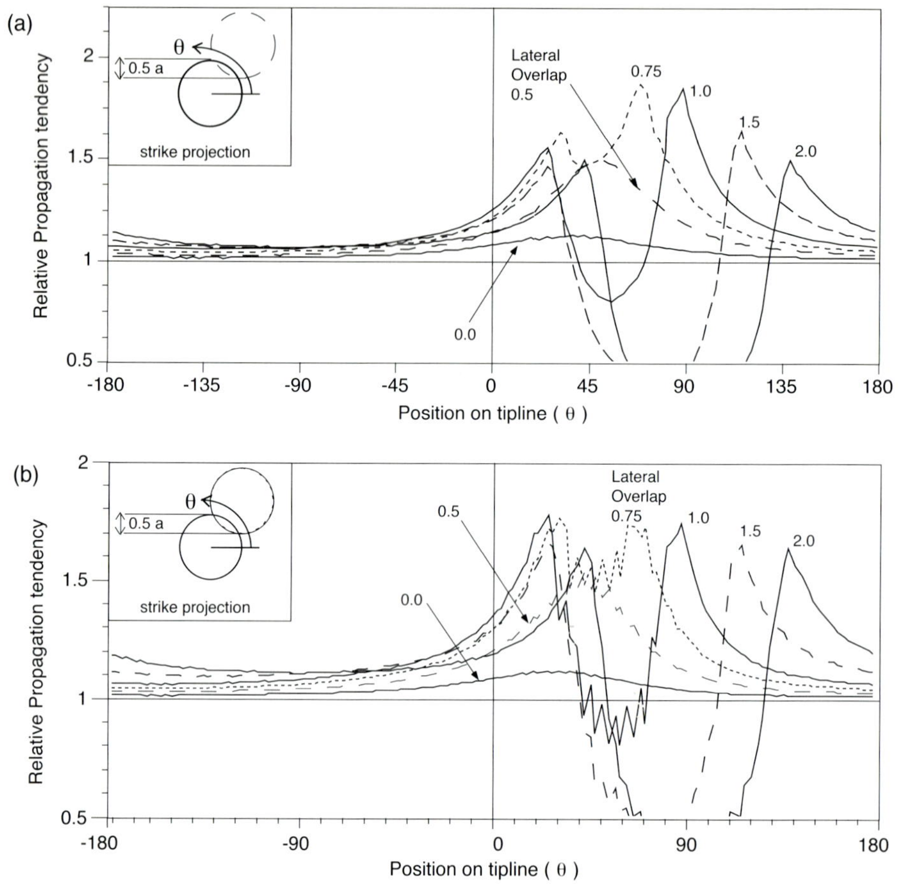

Fig. 10. Propagation tendency along circular faults with a vertical overlap of $0.5a$, for various magnitudes of lateral fault overlap. (a) Variation of propagation tendency along the tipline for a perfectly compressible material (Poisson's ratio identical to zero). Except for the curves for lateral overlaps of $0.5a$ and $2.0a$, the distribution is asymmetric and directly correlates with the spatial fault arrangement. See text for discussion. (b) Variation of propagation tendency along the tipline for material with Poisson's ratio equal to 0.25. The curves are more symmetric because the effect of Poisson's ratio approximately compensates the asymmetric fault arrangement *for these particular spatial fault arrangements*. See text for discussion.

Because of this, initially circular faults should propagate toward one another and may develop an elliptical tipline shape (Fig. 11b). Such underlapping faults would tend to grow until they overlap at their mid-height (Fig. 11c), with the amount of overlap being similar to the fault spacing. At that stage, the fault tips have grown past one another and consequently experience a reduced driving stress at these levels, inhibiting further propagation. At shallower and deeper levels the faults still underlap, and consequently segments of the tipline at these levels experience an increased propagation tendency (Fig. 11c). This arrangement may result in local arrest of the fault tipline in the relay zone, but continued propagation at other levels outside the relay zone where the faults still underlap. In this manner, mechanical interaction would lead to 'straightening' of the fault tiplines in the relay area (Fig. 11d), to become parallel to the slip vector. Such steeply-plunging, relatively straight tiplines are frequently observed, as shown in Figures 2 and 3.

Figure 12 shows a projection of three circular normal fault segments that have obtained the maximum overlap along their mid-height. Slip contours delineate the distribution of total slip on the three segments. It can be seen that very steep slip gradients occur at the top and bottom of the relay zones between the segments. The steep slip gradients are partly caused by mechanical interaction. The local reduction in fault height (as measured parallel to the slip vector) also contributes significantly to the steep slip gradient (Willemse, 1997). Therefore the greatest propagation tendency for the composite fault zone occurs at the upper and lower boundaries of the relay zones, and is directed vertically.

The distribution of total slip in Figure 12 is similar to that of a single, continuous elliptical fault, illustrating that the three segments act kinematically like a single fault. The middle segment has the largest maximum slip, and the slip gradients along the lower and upper part of its tipline are greater than the slip gradients along the tiplines of the two outer segments. This spatial distribution of near-tip fault slip suggests that subsequent fault propagation would take place mostly upwards and downwards in the relay areas and along the middle segment, until a smoother tipline with more uniform slip gradients has been re-established. This is schematically illustrated in Figure 11e. The arrest of the lateral sides of the tipline of the middle segment, and its preferential propagation at the upper and lower parts of the tipline is one possible explanation for the exceptional down dip height (i.e. low aspect ratios) of some fault segments described by Nicol et al. (1996).

The same processes are thought to occur for vertically stepping normal faults. In this case, the spatial alternation of enhanced and inhibited fault propagation is such that straight, horizontal tiplines are formed in relay zones. These tiplines are perpendicular to the slip vector. Nicols et al. (1995) observed such subhorizontal linkage / branch lines. Mansfield et al. (1996) document an example using anomalies in the displacement distribution.

For the general case of faults that step both laterally and vertically, the propagation tendency generally varies asymmetrically along the tipline depending on the spatial arrangement of the faults and, to a lesser extend, Poisson's ratio.

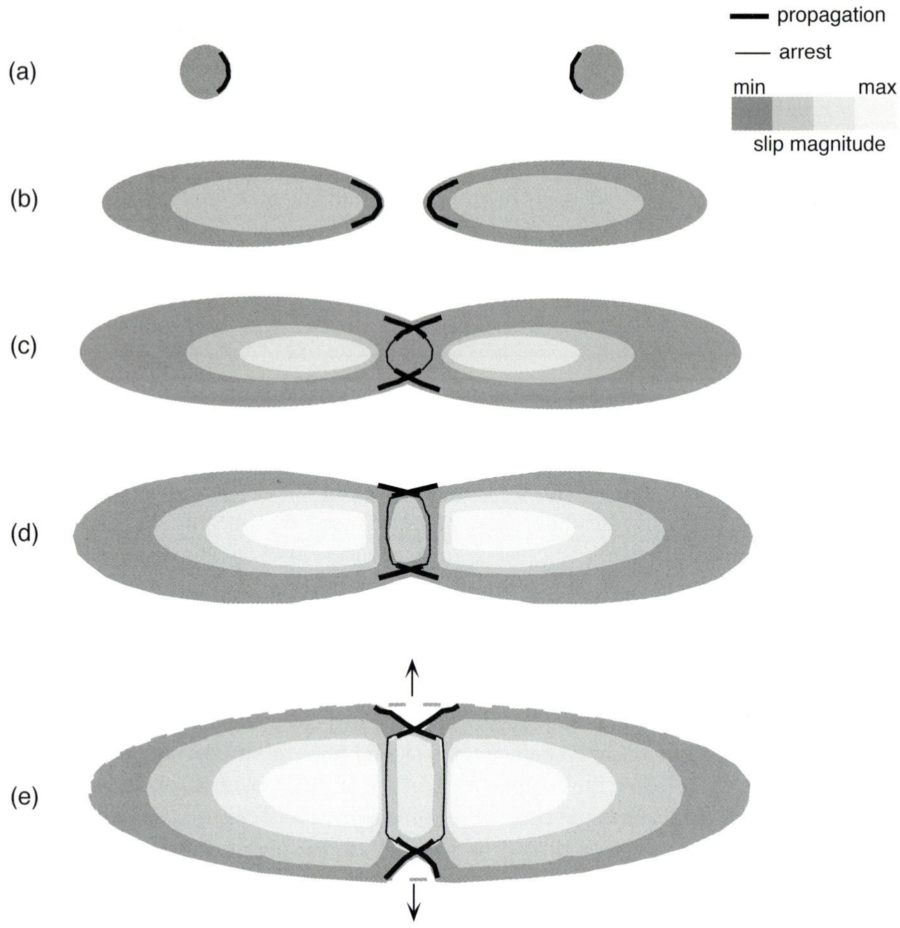

Fig. 11. Conceptual evolution of fault growth and coalescence as shown on a strike projection. Thick lines indicate significantly enhanced propagation tendency; thin lines indicate fault arrest. (a) Initial stage of two circular flaws. The smaller the distance between the tiplines, the greater the propagation tendency . (b) Fault propagation leads to two elliptical, underlapping faults. The faults still have an increased propagation tendency, especially where the distance between them is least. (c) The tiplines of partially overlapping faults are arrested in the region of greatest overlap (thin tiplines), but preferentially propagate immediately above and below this region (thick tiplines). (d) A relay zone is formed that is laterally bounded by steeply plunging, slip-parallel tip lines. (e) The greatest propagation tendency along the composite fault occurs at the top and bottom of the relay, and to the sides of the relay. Continued propagation re-established a smoother tipline

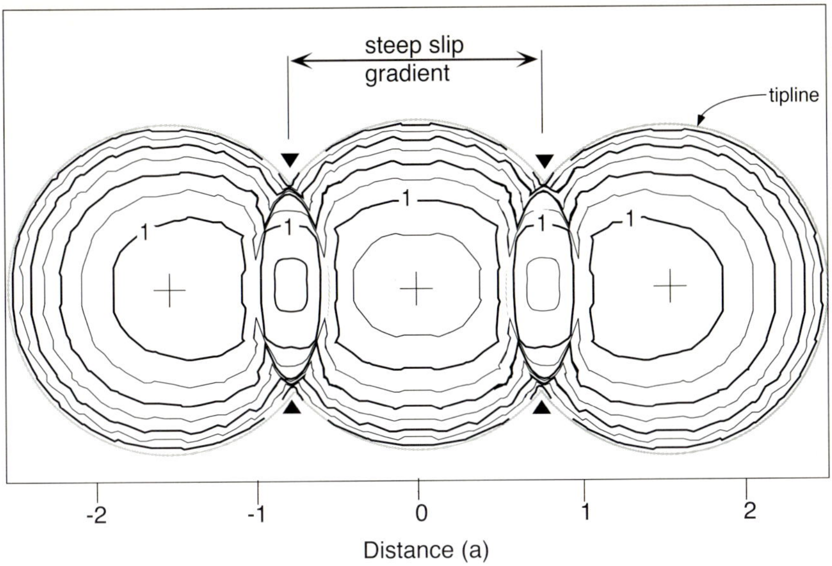

Fig. 12. Contours of total fault slip along array of three circular normal faults. Slip is greatest on the central segment. The greatest near-tip slip occurs at the top and bottom of the relay zones (indicated by solid triangles). Also the top and bottom sections of the tipline of the central segment have comparatively greater slip gradients.

Consequently, propagation is enhanced preferentially either at the upper and lower parts of the tipline, or at the lateral sides of the tiplines. This, together with local regions of fault overlap and hence tipline arrest, could cause faults to propagate either up and down past each other, or laterally past one another until they become wrapped around each other in a manner similar to that shown in Figure 2c. In the region of overlap, the tiplines would tend to be either parallel or perpendicular to the fault slip vector, similar to what is sometimes observed (Fig. 2c, Fig. 4b and Fig. 11 in Nicol et al. 1996). A final geometry may develop where the different fault segments 'fill' the fault zone with a patchwork quilt pattern as shown in Figure 13.

6 Discussion

Our model does not explicitly model fault propagation. Rather, we compute propagation tendency around the tipline of stationary faults and use the results to *infer* how faults might propagate. The model ignores many of the complexities that may play important roles in the development of natural faults. Firstly, the fault model is an idealisation of a single slip event or a series of events that are not accompanied by any stress relaxation. Secondly, the models only address mechanical interaction through perturbation in the stress component σ_{12}, and ignore changes in Coulomb-type frictional resistance at contractional or extensional relay zones. Thirdly, there are other sources for non-elliptical tipline shapes that we have not modeled here. Synsedimentary growth faults, for example, have a straight subhorizontal upper tipline (Nicol et al., 1996). Similarly, exceptionally competent layers may lead to local subhorizontal tipline sections and complex slip distributions. Furthermore, our choice of propagtion criterion restricts the faults to grow in their own planes by simply creating new surfaces of displacement discontinuity. Curvature of the fault plane observed in some relay zones along natural faults is hence not adressed here (Du and Aydin, 1995; Peacock and Sanderson, 1995). Finally, the physical mechanism for the creation of these surfaces is left unspecified. Both laboratory experiments of shear fractures (Peng and Johnson, 1972; Wong, 1986; Petit and Barquins, 1988; Shen et al., 1995) and observations of faults in nature (McGarr et al., 1979; Aydin and Johnson, 1983; Segall and Pollard, 1983; Peacock and Sanderson, 1991) suggest that propagation is a complex process involving a number of different physical mechanisms such as formation and coalescence of micro-cracks, granulation, particulate flow of grains and pressure solution. The mechanisms depend both on the lithology and on the stress state in the near-tip region. Because these stress states are quite different for Mode II and III loading conditions, it is reasonable to conclude that the fault might propagate by different mechanisms at different locations on the tipline, depending on the ratio of sliding to tearing. Our criterion, on the other hand, is based only on the energy release rate, and ignores these possible differences. The energy release rate is calculated as the work done by forces acting at points along the new surfaces of displacement discontinuity, and both the forces and displacements are taken as the near-tip approximations without any further consideration of the actual physical propagation mechanisms.

Despite all these limitations, we believe that our models provide some insight into the evolution of faults that form by coalescence of smaller segments and supply an explanation of some systematic observations along natural faults. For relay zones between laterally stepping normal faults, we have shown that mechanical interaction leads to the formation of steeply plunging, straight tiplines that are parallel to the slip vector. Such straight relay tiplines could make subsequent fault linkage 'efficient' in the sense that fault continuity would be achieved along a continuous line rather than at one or a few points. The slip-parallel attitude of the line of linkage also tends to minimize kinematic problems during subsequent simultaneous slip on the adjacent segments. Such linkage geometries thus may remain intact as slip accumulates and are indeed frequently observed. Mechanical interaction also explains why the downdip dimension of laterally stepping fault segments can be greater than the horizontal segment length.

Vertically stepping normal faults, in contrast, interact to form subhorizontal straight tiplines. Subhorizontal bridges of intact rock between vertically stepping normal faults are perpendicular to the slip vector and hence are likely to be obliterated as slip accummulates. However, relics of breached horizontal relays have been recognized as horizontal zones of abnormal displacement gradients (Mansfield and Cartwright, 1996). Vertically stepping fault segments are likely to have relatively great aspect ratios indicating a comparatively small down dip fault height. Faults that step both laterally and vertically tend to propagate around one another. Such faults also tend to propagate until a limited overlap has been obtained along relay zones that are either parallel or perpendicular to the fault slip vector.

Local interactions across relays between individual faults and the overall slip distribution accommodated by all fault segments tend to promote fault propagation until the envelope surrounding all the individual segments is smooth. The total slip accommodated across the individual fault segments is similar to that of a single, large continuous fault. Thus, the overall effect of mechanical interaction between neighbouring fault segments is to establish order and systematic geometries in terms of (a) segment tipline orientations, (b) the shape of the composite fault zone, and (c) encouraging relatively smooth and simple distributions of total fault slip. In this sense, the sequence of events shown in Figure 11, lateral propagation followed by vertical propagation, is somewhat artificial.

Whereas we only have shown results for uniformly loaded faults, we have also investigated, using (15), the propagation tendency for CEZ faults for a variety of loading conditions and spatial arrangements. The uniformly loaded faults tend to cease propagating when the segment overlap is comparable to their spacing. CEZ fault segments, however, propagate untill the overlap is as much as several times their spacing (Willemse, 1997). Despite such subtle differences, the mechanics of uniformly loaded faults and CEZ faults are very similar. Therefore we believe the above concept for fault growth also to be viable for CEZ faults. This is important because the bell-shaped to near-linear slip distribution along CEZ faults appears a better approximation to many natural faults than the elliptical slip distribution along uniformly loaded faults.

Detailed outcrop studies suggest that faults that consist of a single slip plane generally have small offsets, say less than one metre. Most faults with larger offsets appear to be zones of multiple closely-spaced slip planes. What appears as a single fault surface on a large scale, actually may consist of a complicated patchwork of smaller segments. Each of these segments in turn may have formed by coalescence of even smaller discontinuities, as shown schematically in Figure 13. Such a fault model reflects a progressive process of localization, starting from many small flaws, into fault segments, into a large single composite fault zone. Most of the segments may have coalesced completely, leaving no or only subtle relics of the initial geometry. However, bridges of intact rock between fault segments may be preserved locally on a wide range of scales. Although some bridges may break during fault growth, others are created as new segments become incorporated in the fault.

This model for fault development agrees well with earthquake data. According to some geophysicists, the Gutenberg-Richter frequency - magnitude relation of multiple earthquakes occurring within a certain region and time period suggests that irregularities exist on a fault at all length scales (Andrews, 1980). The high-frequency seismic radiation and strong motions from individual large earthquakes are interpreted in a similar light and believed to be due to an irregular slip

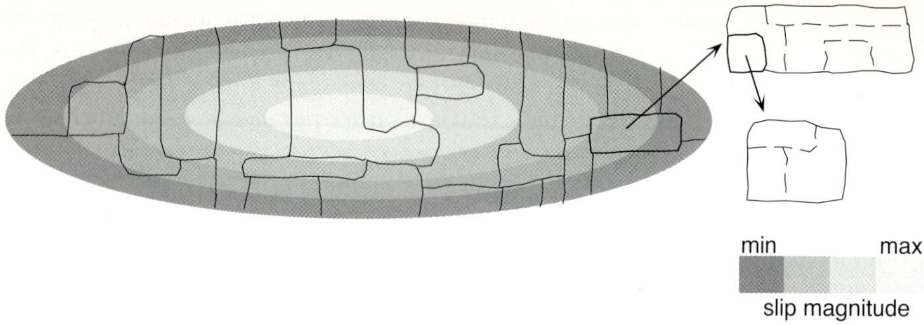

Fig. 13. Conceptual model of single fault. The fault formed by coalescence of smaller segments which themselves formed by coalescence of yet smaller segments. Most of these originally distinct segments are not likely to be preserved with continued accumulation of slip. Locally, however, relays may form bridges of intact rock.

distribution over a heterogeneous fault plane. The "barrier" earthquake model (Das and Aki, 1977) and the "asperity" earthquake model (Kanamori, 1981), both represent this school of thought by postulating strength heterogeneities or geometrical irregularities along the rupture plane. Our fault model provides a physical explanation why natural faults have both strength (intact rock bridges) and geometrical irregularities (relays) at a variety of scales.

The notion that even relatively simple faults may form by segment linkage also has important implications for their hydraulic properties. Particularly important in this respect is the presence of bridges of intact rock that may allow across-fault transport of water or hydrocarbons. The largest intact bridges, that most significantly affect flow at the fluid production time scale of tens of years, may be detectable by careful analysis of the slip distributions. Although most smaller bridges are likely to brake with continued fault growth, the remaining intact ones may be very hard to detect with seismic methods. However, such minor bridges may significantly reduce the fault sealing potential on the time scale of thousands to millions of years most appropriate for migration and retention of hydrocarbons.

Despite the simplicity of our numerical and conceptual models, we find the correspondence to natural normal faults very encouraging. Because the basic relationships between fault geometry, stress drop, rock elastic constants and slip (or opening) are similar (Pollard and Segall, 1987), we expect that very similar changes in segment shape and aspect ratio may accompany coalescence and growth of opening mode cracks such as joints and dikes, and the development of strike-slip and thrust faults.

Acknowledgments E.W. expresses his gratitude to Dr. G. Mandl for the guidance, inspiration and cordial encouragement provided while at KSEPL. His acute analysis of many fracture and fault problems had an exemplary influence on the Structural Geology Group there. Dr. Ing. F.K. Lehner and Prof. E. Wallbrecher are thanked for organising the symposium and providing travel funds. We are grateful to Conrad Childs for kindly reprojecting Figure 3b. Financial support by the Rock Fracture Project at Stanford University and by a Shell Research B.V. (The Netherlands) summer grant to E. Willemse are gratefully acknowledged. Many thanks to Paula, Fer and Marie-Thérèse for support.

References

Aki, K., Characterization of barriers on an earthquake fault, J. Geophys. Res, 84, 6140-6148, 1979.

Andrews, D.J., A Stochastic Fault model: 1. Static case, J. geophys. Res., 85, 3867 - 3877, 1980.

Aydin, A., and A.M. Johnson, Analysis of faulting in porous sandstones, J. Struct. Geol., 5, 19 - 31, 1983.

Aydin, A., and A. Nur, Evolution of pull-apart basins and their scale independence, Tectonics, 1, 91 - 105, 1982.

Aydin, A., and R.A. Schultz, Effect of mechanical interaction on the development of strike-slip faults with echelon patterns, J. Struct. Geol., 12, 123 - 129, 1990.

Barenblatt, G.I., Mathematical theory of equilibrium cracks in brittle fracture, Adv. Appl. Mech., 7, 55 - 129, 1962.

Barnett, J.M., J. Mortimer, J. Rippon, J.J. Walsh, and J. Watterson, Displacement geometry in the volume containing a single normal fault, Am. Assoc. Pet. Geol. Bull., 71, 925-937, 1987.

Becker, A.A., The boundary element method in engineering, 337 pp., McGraw-Hill Book Company, 1992.

Bouvier, J.D., C.H. Kaars-Sijpesteijn, D.F. Kluesner, C.C. Onyejekwe, and R.C. Van Der Pal, Three-Dimensional Seismic Interpretation and Fault Sealing investigations, Nun River Field, Nigeria, Am. Assoc. Pet. Geol. Bull., 73, 1397-1414, 1989.

Bürgmann, R., D.D. Pollard, and S.J. Martel, Slip distributions on faults: effects of stress gradients, inelastic deformation, heterogeneous host-rock stiffness, and fault interaction, J. Struct. Geol., 16, 1675 - 1690, 1994.

Childs, C., S.J. Easton, B.C. Vendeville, M.P.A. Jackson, S.T. Lin, J.J. Walsh, and J. Watterson, Kinematic analysis of faults in a physical model of growth faulting above a viscous salt analogue, Tectonophysics, 228, 313 - 329, 1993.

Childs, D., J. Watterson, and J.J. Walsh, Fault overlap zones within developing normal fault systems, J. Geol. Soc. Lond., 152, 535 - 549, 1995.

Comninou, M.A., and J. Dunders, The angular dislocation in a half-space, J. Elasticity, 5, 203 - 216, 1975. Cottrell, B., and J.R. Rice, Slightly curved or kinked cracks, Int. J. Fract., 16, 155 - 169, 1980.

Cowie, P.A., and C.H. Scholz, Physical explanation for the displacement-length relationship of faults, using a post-yield fracture mechanics model, J. Struct. Geol., 14, 1133 - 1148, 1992.

Cox, S.J.D., and C.H. Scholz, On the formation and growth of faults: an experimental study, J. Struct. Geol., 10, 413 - 430, 1988.

Crouch, S.L., and A.M. Starfield, Boundary element methods in solid mechanics, 322 pp., George Allen and Unwin Ltc, London, 1983.

Cruikshank, K.M., G. Zhao, and A.M. Johnson, Duplex structures connecting fault segments in Entrada Sandstone, J. Struct. Geol., 13, 1185 - 1196, 1991.

Das, S., and K. Aki, Fault Plane with barriers: a versatile earthquake model, J. geophys. Res., 82, 5658-5670, 1977.

Davy, P., On the frequency-length distribution of the San Andreas Fault system, J. Geophys. Res., 98, 12141 - 12151, 1993.

Dawers, N.H., and M.H. Anders, Displacement-length scaling and fault linkage, J. Struct. Geol., 17 (5), 607 - 614, 1995.

Du, Y., and A. Aydin, The maximum distortional strain energy density criterion for shear fracture propagation with applications to the growth paths of en echelon faults, Geophys. Res. Let., 20, 1091 - 1094, 1993.

Du, Y., and A. Aydin, Shear fracture patterns and connectivity at geometric complexities along strike-slip faults, J. geophys. Res., 100, 18093 - 18102, 1995.

Dugdale, D.S., Yielding of steel sheets containing slits, J. Mech. Phys. Solids, 8, 100-104, 1960.

Gay, N.C., and W.D. Ortlepp, Anatomy of a mining-induced fault zone, Geol. Soc. Am. Bull., 90, 47-58, 1979.

Gillespie, P.A., J.J. Walsh, and J. Watterson, Limitations of dimension and displacement data from single faults and the consequences for data analysis and interpretation, J. Struct. Geol., 14, 1157 - 1172, 1992.

Ida, Y., Cohsive force across the tip of longitudinal shear crack and Griffith's specific surface energy, J. Geophys. Res., 77, 3796 - 3805, 1972.

Irwin, G.R., Analyses of stresses and strains near the end of a crack traversing a plate, J. Appl. Mech., 24, 361 - 364, 1957.

Jev, B.I., C.H. Kaars-Sijpesteijn, M.P.A.M. Peters, N.L. Watts, and J.T. Wilkie, Akaso Field, Nigeria: Use of integrated 3-D seismic, fault slicing, clay smearing, and RFT-pressure data on fault trapping and dynamic leakage, Am. Assoc. Pet. Geol. Bull., 77, 1389 - 1404, 1993. Jeyakumaran, M.,

J.W. Rudnicki, and L.M. Keer, Modeling slip zones with triangular dislocation elements, Seis. Soc. Am. Bull., 82, 2153 - 2169, 1992.

Kanamori, H., The nature of seismicity patterns before large earthquakes., in Earthquake prediction: an international review, edited by D. Simpson, and Richards, AGU Maurice Ewing Series, 1981.

Kronberg, P., Geometries of extensional fault systems, observed and mapped on aerial and satellite photographs of Central Afar, (Ethiopia/Djibouti), Geologie en Mijnbouw, 70, 145-161, 1991.

Larsen, P.H., Relay structures in a Lower Permian basement-involved extension system, East Greenland, J. Struct. Geol., 10, 3-8, 1988.

Lawn, B.R., and T.R. Wilshaw, Fracture of Brittle Solids, Cambridge University Press, Cambridge, 1975.

Li, V.C., Mechanics of shear rupture, in Fracture mechanics of rocks, edited by B.K. Atkinson, pp. 351 - 428, Academic Press, London, 1987

Mandl, G., Discontinuous fault zones, J. Struct. Geol., 9, 105-110, 1987.

Mandl, G., Mechanics of Tectonic Faulting: Models and Basic Concepts, 407 pp., Elsevier, Amsterdam, 1988.

Mansfield, C.S., and J.A. Cartwright, High resolution fault displacement mapping from 3-D seismic data: evidence for dip-linkage during fault growth, J. Struct. Geol., 18, 249 - 263, 1996.

McGarr, A., D.D. Pollard, N.C. Gay, and W.D. Ortlepp, Observations and analysis of structures in exhumed mine-induced faults, in Conference VIII: Analysis of actual fault zones in Bedrock., pp. 101 - 120, USGS, Menlo Park, California, 1979.

Moore, D.E., and D.A. Lockner, The role of microcracking in shear-fracture propagation in granite, J. Struct. Geol., 17, 95-114, 1995.

Morley, C.K., R.A. Nelson, T.L. Patton, and S.G. Munn, Transfer zones in the East African rift system and their relevance to hycrocarbon exploration in rifts, Am. Assoc. Pet. Geol. Bull., 74, 1234-1253, 1990.

Muraoka, H., and H. Kamata, Displacement distribution along minor fault traces, J. Struct. Geol, 5, 483-495, 1983

Nicol, A., J.J. Walsh, J. Watterson, and P.G. Bretan, Three-dimensional geometry and growth of conjugate normal faults, J. Struct. Geol., 17 (6), 847 - 862, 1995.

Nicol, A., J. Watterson, J.J. Walsh, and C. Childs, The shapes, major axis orientations and displacement patterns of fault surfaces, J. Struct. Geol., 18, 235 - 248, 1996.

Olson, J., Fracture Mechanics analysis of joints and veins, PhD thesis, Stanford University, 1991.

Olson, J.E., and D.D. Pollard, Inferring stress states from detailed joint geometry, in Key questions in Rock Mechanics, edited by P.A. Cundall, R.L. Sterling, and A.M. Starfield, pp. 159 - 167, Balkema, Rotterdam, 1988.

Olson, J.E., and D.D. Pollard, The initiation and growth of en echelon veins, J. Struct. Geol., 13, 595 - 608, 1991.

Palmer, A.C., and J.R. Rice, The growth of slip surfaces in the progressive failure of overconsolidated clay, Proc. Roy. Soc. Lond., A332, 527 - 548, 1973.

Paris, P.C., and G.C. Sih, Stress analysis of cracks, in Fracture Toughness Testing, pp. 30-83, American Society for Testing Materials, Special Technical Publication, 1965.

Peacock, D.C.P., Displacement and segment linkage in strike-slip fault zones, J. Struct. Geol., 13, 1025 - 1035, 1991.

Peacock, D.C.P., and D.J. Sanderson, Displacements, segment linkage and relay ramps in normal fault zones, J. Struct. Geol., 13, 721-733, 1991.

Peacock, D.C.P., and D.J. Sanderson, Geometry and development of relay ramps in normal fault systems, Am. Assoc. Pet. Geol. Bull., 78, 147 -165, 1994.

Peacock, D.C.P., and D.J. Sanderson, Strike-slip relay ramps, J. Struct. Geol., 17, 1351 - 1360, 1995

Peng, S., and A.M. Johnson, Crack growth and faulting in cylindrical specimens of Chelmsford granite, Int. J. Rock Mech. and Mining Sci., 9, 37 - 86, 1972.

Petit, J.P., and M. Barquins, Can natural faults propagate under mode-II conditions?, Tectonics, 7, 1243 - 1256, 1988.

Pollard, D.D., and A. Aydin, Propagation and linkage of oceanic ridge segments, J. Geophys. Res., 89, 10017 - 10028, 1984.

Pollard, D.D., S.D. Saltzer, and A.M. Rubin, Stress inversion methods; are they based on faulty assumptions?, J. Struct. Geol., 15, 1045-1054, 1993.

Pollard, D.D., and P. Segall, Theoretical displacements and stresses near fractures in rock: with applications to faults, joints, veins,dikes, and solution surfaces, in Fracture mechanics of rocks, edited by B.K. Atkinson, Academic Press, London, 1987.

Pollard, D.D., P. Segall, and P. Delaney, Formation and interpretation of dilatant echelon cracks, Bull. Geol. Soc. Am., 93, 1291 - 1303, 1982.

Rice, J.R., Mathematical analysis in the mechanics of fracture, in Fracture: An advanced treatise, edited by H. Liebowitz, pp. 191 - 311, Academic Press, San Diego, 1968.

Rice, J.R., Theory of precursory processes in the inception of earthquake rupture, Gerlands Beitrage Geophys., 88, 91 - 127, 1979.

Rippon, J.H., Contoured patterns of throw and hade of normal faults in the coal measures (Westphalian) of northeast Derbyshire, Proc. Yorks. Geol. Soc., 45, 147-161, 1985.

Rudnicki, J.W., Fracture mechanics applied to the earth's crust, Ann. Rev. Earth Planet. Sci., 8, 489 - 525, 1980. Segall, P., and D.D.

Pollard, Mechanics of discontinuous faults, J. Geophys. Res., 85 (B8), 4337 - 4350, 1980.

Segall, P., and D.D. Pollard, Nucleation and growth of strike slip faults in granite, J. Geophys. Res., 88, 555 - 568, 1983.

Shen, B., O. Stephansson, H.H. Einstein, and B. Ghahreman, Coalescence of fractures under shear stresses in experiments, J. Geophys. Res., 100 (B4), 5975 - 5990, 1995.

Sih, G.C., Three Dimensional Crack Problems, 452 pp., Noordhoff International, Leiden, 1975.

Sneddon, I.N., The distribution of stress in the neighbourhood of a crack in an elastic solid, Proc. Roy. Soc. London, Section A, 187, 229 - 260, 1946.

Tada, H., P.C. Paris, and G.R. Irwin, The Stress Analysis of Cracks Handbook, Paris Productions Incorporated (and Del Research Corporation), St. Louis, 1985.

Thomas, A.L., Poly3D : a three-dimensional, polygonal element, displacement discontinuity boundary element computer program with applications to fractures, faults, and cavities in the Earth's crust, MSc thesis, Stanford University, 1993.

Thomas, A.L., and D.D. Pollard, The geometry of echelon fractures in rock: implications from laboratory and numerical experiments, J. Struct. Geol., 15, 323 - 334, 1993.

Walsh, J.J., and J. Watterson, Analysis of the relationship bewteen displacements and dimensions of faults, J. Struct. Geol., 10, 239 - 247, 1988.

Walsh, J.J., and J. Watterson, Displacement gradients of fault surfaces, J. Struct. Geol., 11 (3), 307 - 316, 1989.

Walsh, J.J., and J. Watterson, Geometric and kinematic coherence and scale effects in normal fault systems, in The Geometry of Normal Faults, edited by A.M. Roberts, G. Yielding, and B. Freeman, pp. 193 - 203, Geol. Soc. Lond. Spec. Publ., 1991.

Willemse, E.J.M., Segmented normal faults: correspondence between three-dimensional mechanical models and field data, J. Geophys. Res., 102, 675 - 692, 1997.

Willemse, E.J.M., D.D. Pollard, and A. Aydin, 3D analysis of slip distributions along echelon normal fault arrays with consequences for fault scaling, J. Struct. Geol., 18, 295 - 309, 1996.

Williams, M.L., On the stress distribution at the base of a stationary crack, J. Appl. Mech., 24 (109 - 114), 1957.

Wong, T.F., On the normal stress dependency of the shear fracture energy, in Earthquake source mechanics, edited by S. Das, J. Boatwright, and C.H. Scholz, pp. 1-12, AGU, Washingon, D.C., 1986.

Printing (Computer to Film): Saladruck, Berlin
Binding: Stürtz AG, Würzburg